HOW TO

DEVELOP

THE

MOON

Ian Long

Dad, Jennifer, Stephen, Veronica…
-Thanks

Composition of Lunar and Terrestrial Crust

Element	Earth	Moon
O (wt%)	46.0	41.6-44.6
Si (wt%)	28.8	19.8-21
Fe (wt%)	4.3	4-12.3
Al (wt%)	8.0	7.3-14.4
Ca (wt%)	3.9	7.5-11.3
Mg (wt%)	2.2	3.5-6
Ti (wt%)	0.4	0.3-4.6
Na (ppm)	23,600	2000-5000
K (ppm)	21,400	1000
P (ppm)	98	800
Mn (ppm)	716	200-2000
Ni (ppm)	56	200
Zr (ppm)	203	100-400
V (ppm)	98	130
C (ppm)	200-1990	<100
F (ppm)	525	70
Cl (ppm)	472	50
Co (ppm)	24	40
Li (ppm)	18	10
Cu (ppm)	25	8

From Apollo 11, 12, and 16. Lunar regolith data is extremely limited and should be taken with a grain of sodium ions.

CONTENTS

WHY?   1

A COLD START   8

ADDRESSING THE XLEPHANT   13

STELLAR STEPS   19

STAINLESS STEEL SEEDS   23

I TRISO HARD BUT IN THE END   34

CHROMIUM ANODES   47

BREAK GROUND   56

FOOD & WATER   60

THE ENVELOPE'S BACK   68

PROFIT   74

WALMART   82

INFRASTRUCTURE   93

THE ENTROPIC ABYSS   104

WHY?

Today detractors argue that "*space colonization*" is about greed, ego, and ambition; that it stems from the rotten human urge to *conquer* everything in sight. They are wrong. It is not about conquest. It's about carrying capacity.

Think of your proudest moment, the times when you felt a surge of love, or witnessed the miraculous complexities of nature — all these echoes in the grand symphony of life. Nuclear war is particularly terrifying because it would mean your entire life, all those echoes, were pointless. And not just your life– but all human lives including those who came before: everything literally everyone did, all the effort, struggle, and sacrifice to get to this point– ALL OF HISTORY– every time a mother worried, every time a father was proud, every achievement, every failure… all would be rendered meaningless if the Earth is wasted in Armageddon.
In this final act the ultimate question would be answered: *what's the point?*

Conclusively, it would appear, there was none.
Thus nuclear annihilation is the ultimate form of nihilism proving every suicidal thought correct. Every triumph over pain and suffering was meaningless.

It was going to end painfully anyway.

It gets worse. There are more dangers lurking in the dark. Nuclear war is just one potential future fail-state. The list of mass extinction events throughout Earth's history is long but five stand out as particularly brutal. Five times over the past 3.7 billion years of history where biomass fell so far that all known life in the universe teetered perilously at the brink of extinction. To give you a sense, the Permian–Triassic also known as 'The Great Dying' was Earth's largest extinction event and saw the obliteration of 57% of biological families, 83% of genera, 81% of marine species, and 70% of terrestrial vertebrate species. It was caused by a relentless period of massive volcano eruptions, global warming, and asteroid impacts. It took *30 million years* for the terrestrial vertebrate fauna to fully recover. This is but one of four events that serve as stark reminders of the thin line that separates Earth's vibrancy from void, existence from extinction, and consciousness from cosmic oblivion. Nuclear war might happen. There's potential. But it has never happened before. And, fortunately, it is entirely within our control. With nuclear war, we would be the authors of our own destruction. Not so with these others…. These, unfortunately, have happened before. Entirely out of our control. Events like these will happen again. It's only a matter of time.

Impacts by comets and asteroids are obvious. Less so is the likely possibility of a near-Earth supernova– a massive stellar explosion obliterating everything within a 1000-light-year radius. At present, there are six near-Earth supernova candidates within this distance.

If that doesn't get us then in about 250 million years all continents will converge to form Earth's next supercontinent, Pangea Ultima in which only about 8% of Earth's land will be habitable for complex life.

The majority of this continent will be desert.
This will seriously impact the real estate market.
You hear that Michael Burry
This will arise as a result of two processes:

First, more and more CO_2 will be released over time as a natural consequence of plate tectonics. By 250 million years the CO_2 in the atmosphere will increase to over five times today's average due to changes in volcanic rifting and outgassing. This will increase the temperature of the planet.

Second, over time the luminosity of the Sun will steadily increase resulting in a rise in the solar radiation reaching Earth. In 250 million years emissions from the sun will have increased by 2.5%. This may not sound like much but combined with the increase in CO_2 levels it's enough to dramatically shift the conditions on the Earth's surface, altering the global climate and resulting in unbearable conditions for complex life.

By this point the Earth's mean temperature is expected to be approximately 46.5 C or 116 F. In comparison the mean temperature on Earth today is only 15 C or 59 F. At these temperatures, most complex modern life would have a lot of trouble surviving as the vast majority of proteins would denature. Temperatures would be more habitable around the poles and in the ocean, thus leaving only about 8% of Earth's landmass habitable to complex life.

In about one billion years, the solar luminosity will be 10% higher, causing the atmosphere to become a "moist greenhouse" boiling off the oceans. So, best case scenario life on Earth has about one billion years left. Life has been on Earth for 3.7 billion years so it's got about 20% of that time left before it's game over, lights out. To put this into perspective let's personify all known life into a single person.

A US male. The average life expectancy for a US male is 73.5 years old. If all known life in the universe was a US male then it's about 59 years old and I'm a doctor telling you that if you don't change your diet by developing the Moon then best case scenario, if everything goes perfectly, you will live for a maximum of 14 and a half years. Better start working on that will.

We are not a young planet and intelligent life only evolved once. *Now.* It took 3.7 billion years to do so. But life doesn't develop the Moon, not even complex life. Humans are the only animals that can actually leave Earth, but we don't do it by ourselves. We built a civilization that makes rockets and space suits and this took time to grow. It took two million years of hominin evolution for modern homo sapiens to emerge, 150 thousand years for the neolithic revolution to occur giving rise to civilization, and 12 thousand years of growth and innovation for space flight to occur. Only in the past 62 years have we been able to leave Earth.

How long can this last?

When people think about the modern world ending they often think something along the lines of:

"Well it grew once, it can grow again. It's unlikely all humans will go extinct in the case of a cataclysm and while it took 160k years for them to unlock spaceflight, this is actually only 0.064% of our 250 million-year deadline– so we actually have lots of time to reinvent civilization over and over again. That's many chances to develop the Moon. Maybe it won't be humans, but some form of intelligent life that comes after. We'll figure it out eventually."

I wouldn't be so sure. It's my contention that even a basic civilization, and especially one that becomes capable of reaching orbit, is not guaranteed to emerge just because you have some intelligent life crawling around. Neanderthals and Denisovans were basically as intelligent as modern humans and they went extinct after 350,000 and 400,000 years respectively without ever building a city. Modern humans have only been around half as long and 93.75% was spent doing nothing particularly special. Even if civilization does rise again it isn't guaranteed to become spacefaring. When civilization finally did emerge it spent the first twelve thousand years just rising and falling, with a highwater mark generally measured by a pyramid. As far as we know there is only one planet capable of harboring life, and life only evolved one time. It took billions of years for complex intelligent life to emerge and out of all complex intelligent life only one species ever built a civilization, and out of all civilization's long history of rising and falling only once did it rise higher than a pyramid– high enough to reach the Moon.

So no, reaching orbit is not a natural conclusion of the march of civilization, and civilization isn't guaranteed to emerge from intelligent life, thus it's not obvious we actually have lots of time to reinvent civilization over and over again.

How long will this weird high water mark last?
How high can this tower of Babel climb?
We truly are at a unique moment.
It's not guaranteed to last.

The window of opportunity is closing and it may never open again. Remember it took 30 million years for the terrestrial vertebrate fauna to fully recover from the Permian–Triassic extinction event. If we care at all about life then we need to do everything within our power to pry that window open.

Life on Earth will be wiped out at some point.
It's guaranteed.

But death is not so scary when your life has meaning derived from knowing that your actions in life will, in some big or small way, help those who come after you. It's easier to die when you have happy healthy children because in some ways you're not really dying, you live on through them as a memory… a ghost. Your life had meaning– you served as a seed from which the future grew. Your life kept the music going in the continuous chaos-reducing, order-producing, information-propagating chemical ballet. Earth's death is not so scary when its life had meaning derived from knowing that its existence allowed life to flourish to the point it was able to propagate across the heavens. It's easier to die when you have happy healthy children prospering across vast swaths of space because in some ways you're not really dying, you live on through them as a memory… a ghost.

Earth's life had meaning– she served as a seed from which the future grew. Earth's life kept the music going in the continuous chaos-reducing, order-producing, information-propagating chemical ballet.

We must become a spacefaring multi-planetary species.
This is how we avoid nihilistic annihilation.
This is how we author a good evolutionary destiny.

To become a spacefaring multi-planetary species, propagating life across the sky, illuminating the entropic abyss with the light of consciousness... creating cloud cities on Venus, manufacturing methane on Mars, mining asteroids for ice and metals, germinating seeds on Ganymede then we must first develop the Moon. Considering humans are the only thing the Earth has ever grown that can actually reach orbit. The burden of responsibility falls on us. We cannot count on aliens or future humans to do it for us, the chances are too slim. Armageddon is waiting for us tomorrow, we have to make the leap now. Within our lifetimes.

So why should you care about developing the Moon? Because every day the Moon is left untouched is a step closer to the nihilistic abyss. The time has come to seriously envisage a settlement blueprint that transcends our earthly boundaries, to plant seeds of life where once there was barrenness. If we don't get this right then in all likelihood we'll be confined to Earth forever. What could have been Humanity's beautiful cradle will become a gilded tomb. Consciousness smothered in the womb.

It's time to develop the Moon.

A COLD START

Man harnessed the power of the atom on July 16th, 1945. World War 2 ended shortly after. When the ash settled former allies, ideologically opposed, found themselves staring down the barrel of each other's war machines, but the fear of atomic heat froze each, so open hostilities remained cold and an iron curtain rose. Thus the world entered into a new era. Things weren't great, but they were better than they had been over the past three decades which were marked by an economic depression sandwiched between two world wars. The wars were over, the economy was recovering, the old colonial system was unraveling, global literacy rates, life expectancy, and social mobility were all increasing, and infant mortality and world hunger were decreasing. In this new era, the future seemed bright. All one had to fear was fear itself. Oh...and *absolute nuclear annihilation.* On the 29th of August, 1949 the Soviet Union secretly conducted its first successful nuclear weapon test.

If the Soviets were good at anything it was ideologically-driven paranoid genocidal purges... and *science.* The Soviets believed that anybody, no matter their race, religion, gender, or social class– all could be made into good communists, they just had to be educated. And educate they did. The Soviets excelled in education, as long as that education agreed with communist ideology, and given that science is typically apolitical and yields great advancements that might give the Soviet war machine a technological edge, science was heavily emphasized. While American kids were competing on football fields and bullying math nerds, Soviet children were competing over chess boards and sending math nerds to Moscow to compete in prestigious state-sponsored tournaments.

Seven years on since the test of the first Soviet atom bomb and Russia–not the US–maintained the technological edge. If there were any who doubted this, the launch of humanity's first satellite by the soviet union on the 4th of October, 1957 quickly dispelled all doubt. We often fail to remember that Sputnik was a satellite purposefully built from a hollowed-out nuclear warhead. But to those on the ground in the US scouring the sky on that early October night, the message was clear.

Terror ensued.

The United States prided itself on being at the forefront of technology and immediately began developing a response. In December of the same year, only three months after the launch of Sputnik, the US attempted to launch a satellite of its own.

It blew up on the launch pad.

But a month later, on January 31st, 1958, the Americans succeeded with the successful launch of Explorer I, the first U.S. satellite to successfully orbit the Earth. In July of that same year, Congress passed legislation officially establishing NASA. The race was on.

12 years later the Apollo program proved the US's ICBM technology was superior to that of the USSR's, but the program was expensive. Like really, really expensive. After six landings it became apparent there were diminishing returns for research value with each mission. Only so many rocks and soil samples were needed and with the costs so high per mission it became less and less attractive to Congress which really preferred to spend US taxpayers' hard-earned dollars prolonging the Vietnam War to fulfill their terms of employment as fairly-elected representatives of the military-industrial complex.

To NASA this wasn't great news– but it was okay as they already had a lot of rocks and soil samples to examine, and it was obvious to everyone that the next step was to establish a permanent lunar outpost. But before that could be done two things needed to happen. First launch costs needed to be reduced, and second it wasn't actually obvious humans could survive in low-gravity high-radiation environments for extended periods of time. Besides the obvious risk of developing cancer, there were concerns about brain and organ deformation, and bone and muscle atrophy, so while some guys worked on tackling the launch cost problem with the development of a reusable launch vehicle (the space shuttle), others worked on the human-body-in-space problem with the simultaneous development of the Skylab space station, launched in 1973.

Over the next 40 years, NASA built a bunch of space shuttles and space stations– all the while being pummeled by budgetary cuts and politics slowing them down every step of the way– but by the time the International Space Station was complete, there was no doubt humans could live in space for prolonged periods of time with only a slight increase in cancer risk, eyeball deformation, and muscular atrophy (given proper exercise). It was time to begin planning a return to the Moon. While NASA had successfully answered the 'human-body-in-space problem' they had been unsuccessful in lowering launch costs. By 2011 the cost to launch a kilogram into orbit had *not fallen at all* since 1969 and the shuttle program, 40 years later, still cost around the same as the Saturn V despite being reusable! Oh well, who cares, it's just money, right? Why should the US Congress care about how many pieces of paper it takes to get into orbit?

So on July 8th, 2011 the last space shuttle flew and two months later the Space Launch System (SLS) project was initiated to develop a new lunar launch vehicle for a new era and a new generation (nevermind the fact it's mostly cobbled together from old hardware made in an old era by an old generation). It needed to be political-proof as NASA was afraid the project might be axed any time Congress reviewed its annual budget; so NASA went straight to the source and contracted the SLS to be built by the same industrial entities that gave birth to the horrific monstrosities lurking in the shadows of Capitol Hill. With their masters making money from SLS no four-year congressional puppet would terminate the project on a whim. Unfortunately, these contracts used the cost-plus contracting funding scheme which works like this: NASA says "*here's how much money we're going to give you to build this rocket, but we'll also pick up the bill on anything that goes over that budget.*" The problem with this scheme is it offers no incentive to remain within budget or on schedule. In fact, timeline slips literally mean more money for the contractors which is why it's the government's favorite method of procurement. This is how we got an SLS six years late and ten billion dollars over budget in 2022. The prime contractor receiving the most money for SLS?

Boeing.

Although NASA does performance reviews of their contractors they've been scrutinized for being way too easy on… well, take a wild guess. Anyway, two years into SLS's development the US government and NASA were being ridiculed publicly for paying the Russians to launch US astronauts and crew to the ISS which was seen as a national embarrassment. It was. But it was also supposed to be a temporary solution during the transition from the shuttle to SLS. Unfortunately, it was also becoming obvious that the SLS was years away from reaching the 2016 deadline and was projected to cost more than both the Saturn V and the Space Shuttle! Around the same time in 2013, this little space startup company was beginning to gain the public's attention by promising reduced launch costs through reusability gained from controlled vertical landings of the first stage. Eight years later NASA chose SpaceX's Starship to be the official Artemis human landing system.

Until recently, the (literally) astronomical costs associated with space launches rendered the notion of lunar development economically unsustainable. Financially unviable. Politically unadvisable. Practically unthinkable. Activities extending beyond orbit, with the exception of telecommunications, were confined to government-led initiatives, characterized by significant bureaucratic red tape, political uncertainty, and reliance on public funding. This has changed. This book will attempt to describe in as much detail as possible how to develop the Moon *profitably*.

ADDRESSING THE XLEPHANT

This is not a book about SpaceX but we can't talk about lunar development without discussing the vertically integrated 394-foot 16.7-million-pound thrust-producing stainless steel water tower in the room. My bias: I like SpaceX. You might not. That doesn't matter, you can't ignore their results. SpaceX has breathed much-needed life into a stagnant industry once heralded as the epitome of innovation and national prestige. Is SpaceX so *good* or is NASA *so bad* that even a modicum of proficiency in this antiquated industry looks like genius? The contrast exists and is so extreme that it has led many space-enthusiast armchair observers to argue that we have not been back to the Moon because NASA is now just another bloated government bureaucracy that exists as a relic from a glorious past now calcified and crushed under a mountain of paperwork wrapped in red tape and budgetary reports.

And... they're kinda right. But what they miss is that NASA is the biggest investor in SpaceX because the *people* who work at NASA are no dummies, and these not-so-dumb people might work at an agency that *literally puts people's heads in the clouds,* but to get some head in the clouds one (ironically) has to be pretty grounded. And on the ground, the private sector does things much faster and cheaper than the public sector because the private sector is more efficient. This efficiency, however, means the private sector is not a good incubator for early-stage technologies that are worth developing but do not yet have a proven track to profitability.

To address this gap, NASA has long maintained a technology transfer program, which facilitates the transition of mature technologies from the public to the private sector. And the extent to which this program has served the US economy is immeasurable. Oh wait, actually, it is measurable: during the fiscal year 2019, the agency generated more than $64.3 billion in total economic output. Its federal budget in 2019 was $21.5 billion, by the way, so it literally *tripled* every dollar thrown into it *in a single year*.

NASA doesn't accomplish this impressive return on investment by just hatching incubated technology into the private sector– it also levels up the workforce by giving professionals experience points through contracting. NASA actually doesn't have that many employees (relatively). Instead, NASA largely contracts employees of other companies to work at NASA for a time on a certain project. This is by design: when private American companies make progress in air and space, that isn't competition for NASA, it's success. That's what the armchair NASA naysayers get wrong. Both those who say NASA is ineffective and outdated or that *only* NASA should develop the Moon fail to understand its intended function. Love it or hate it NASA is a publicly funded institution that serves the public by enhancing the private sector. Ideology aside, on the ground, this is what works because it plays to the strengths of both. It's not flawless but it won the Cold War.

Now SpaceX builds better ships faster and cheaper than anyone else– ships that are paid for in large part by NASA. SpaceX has lowered the astronomical costs associated with space launches that once rendered the notion of lunar development economically unsustainable. We are now on track for it to become financially viable. Politically advisable. Practically thinkable. Activities extending beyond orbit are no longer confined to government-led initiatives, characterized by significant bureaucratic red tape, political uncertainty, and reliance on public funding; and because of that SpaceX now has competition.

Convergent evolution refers to the phenomenon in which distantly related organisms independently evolve similar traits to adapt to similar necessities. In other words, when faced with a problem, natural selection will often produce remarkably similar solutions even in animals separated by millions of years of evolution. Examples: Bats were once rats and birds were feathered lizards but somewhere along the way both learned to fly using very similar physiological structures (wings). Bats fly and dolphins swim but both use echolocation. Dolphins were once dogs (I'm not kidding, google 'pakicetus') and sharks are fish but both have fins and flippers. Convergent evolution is also why so many different species look so similar, king and coral snakes, viceroy and monarch butterflies, wasps and bees, and lizards in trees.

Bat and bird wings are not the same. Bat's wings are skin stretched between a wide scaffolding of bone while bird's wings are feathers protruding from an 'arm' in which the bones are not stretched wide. These physiological structures are different but these differences don't change the underlying physical principles that make them work. Same with dolphin and shark fins. One has bones, the other cartilage. Different evolutionary design journeys, same (basic) results. Which is better? That's not the point– they are both 'good enough' to be competitive. Anything that is not competitive dies off because natural selection lacks government subsidies.

Why did you just read two paragraphs on evolutionary biology in a book about lunar development? The point is we live in a universe with hard physical limits. There is a *right* way and a *wrong* way to do something. You can *try* to compete in the same ecological niche as dolphins but if you lack fins, flippers, and a streamlined body you're not gonna go far. You can *try* to build a super heavy-lift launch vehicle for lunar development but if you lack reusability, economic scale, vertical integration, and full-flow staged combustion cycle engines you're not gonna go far.

SpaceX has competition but as of this writing, none are in the same league. Yet. Right now all of the launch vehicle providers (rocket companies) that aren't SpaceX are basically trying to catch up to where SpaceX was in 2015 and they are all trying to distinguish themselves with one or two headline-catching unique advantages: Electric Rockets! 3D Printed Rockets! Carbon Fiber Rockets! Million Dollar Lobbyist Backed Government Contract-Funded We Can't Fail Because We're The Elite Legacy Technology Rockets! (Now invest please!) But over time as the industry progresses a convergent effect will take hold. We see the same effect with pop music and car manufacturing. They aren't the *exact same* but... well you get the point. These launch vehicle providers may have different "*design journeys*" but in the end, if they are to survive i.e. be a competitive player in lunar development their final product will need to have similar capabilities and cost to current SpaceX technology.

SpaceX's Starship has already been chosen to be NASA's lunar lander, and it currently makes the most sense to be the first lunar outpost. SpaceX has a lead now but others will grow into their own. Market competition is productive and necessary for sustainability. So to be fair to all potential lunar developers (and to avoid this book being mistaken as SpaceX fan fiction) I am going to write in more generic terms. We can't develop the Moon without Starship... or a Starship-*like* launch vehicle; so going forward instead of "*Starship*" we'll say "*Lunarship*" and understand this implies a Starship-like rocket that could be made by anyone. The industry term is *super heavy-lift launch vehicle* but that's a mouthful. Now the golden question: what is the cost to deliver a 200 ton payload to the lunar surface using one of these definitely-not-Starships? As of writing nobody knows for sure but we can guess.

Fortunately we don't have to guess because the people at Payload Research have guessed really educatedly for us in a 36 page report which breaks down Starship's costs piece by piece and explains both how much Starship costs to build and launch as well as how much SpaceX might charge for it. And their conclusions are: $90 million falling to $35 million after several years. This is the base cost and does not include R&D amortization which will increase the cost of the initial Starships to hundreds of millions. But those initial Starships, basically prototypes, will be used by SpaceX to build out their Starlink constellation thus the costs will mostly be internal. Additionally the more a ship is used the lower the cost per flight (a ship that costs $100 million dollars to build but flies 10 times would reduce its build cost to $10 million per flight). The report also estimates a $10 million flight cost for fuel, inspections, amortized costs, refurbishment, and other miscellaneous items like ground control operations. The report estimates a 5 use lifespan for each Starship although this is *highly conservative* and it is likely to be much higher. Currently Starship has an internal fairing volume of 1000 cubic meters and can deliver 200 tons to Low Earth Orbit (LEO) but SpaceX plans to extend the size of the fairing in future ships to reach 300 tons to LEO. Regardless for the remainder of this book we will use:

$50 million build cost.
$10 million flight cost to LEO.
5 use lifespan.
200 ton 1000 cubic meters payload capacity.

The issue, however, is we are not talking about developing low earth orbit, we are talking about developing the Moon and thus the flight cost will be higher. Currently once such a large and heavy rocket reaches low Earth orbit it has exhausted most of its fuel. SpaceX's solution to this is to set up an orbital refueling depot, basically a Starship retrofitted to be a tanker that remains in orbit and receives the leftover fuel from 5-8 launches until its tanks are full. Then any ship designated to carry people and cargo to the Moon or elsewhere launches into orbit and rendezvous with the full tanker which then transfers its fuel into the lunar-bound ship giving it enough juice to get to its destination and back. So for each launch to the Moon 5 to 8 fuel-launches and orbital docking sequences have to be performed. If we take the higher number (8) and each of those cost $10 million then our flight cost to the Moon *and back* is $80 million.

$50 million build cost.
$80 million flight cost to the Moon.
5 use lifespan.
200 ton 1000 cubic meters payload capacity.

Per flight cost: $90 million.
Cost per kilogram to lunar surface: $450.
Let's round up to $100 million and $500 per kilogram.

STELLAR STEPS

For this book, we will define the steps of lunar development as survey missions, outposts, bases, settlements, and, eventually, a colony. Survey missions come first. These are your intrepid explorers venturing into the unknown. Christopher Columbus, James Cook, Henry Hudson, Lewis and Clark, Nikolay Przhevalsky, Ernest Shackleton, Neil Armstrong. Survey missions establish the viability of a venture and test technology. It's a misconception that people believed the world to be flat before Columbus. 15th-century Europeans knew geometry and had accurate approximations of how large the world was (the Greeks had worked this out in 240 B.C.). Remember Columbus was searching for the Orient, and he was rejected three times over seven years before getting funded by the Spanish monarchs not because people were 'close-minded' but because they could just do the math and see that it was actually further to sail westward to the orient than sailing South and then East around Africa. This is why nobody had tried it before. It seemed dumb. Before Columbus, it was generally believed there was a giant ocean between Europe and Asia (of which there are two). Columbus didn't show everyone the world was round, but he did show there was not a giant ocean between Europe and Asia. More importantly, his expeditions showed that through the use of cutting-edge technology, with the Caravels' shallow draught and lateen sails, one could not just get to the New World but actually return from it.

Before the Antarctic expeditions, it was generally believed Antarctica might be a collection of islands rather than a singular continent, and before the Apollo missions, it was feared there was real concern that the lunar landers would sink deep into the surface as the lunar surface material was thought to be so 'fluffy' it might swallow the lander whole. The Apollo missions were the first survey missions to the

Moon and the upcoming Artemis missions will continue the work. They will show the viability of modern technology in being able to sustain long-term lunar dwellings.

Outposts come next. Think of the first structures in Antarctica. The International Space Station (ISS) can be considered a research outpost. Outposts are typically in very remote and inhospitable places where research is typically conducted. This is usually done by a small team of highly trained experts who will remain in these locations for a number of weeks, months, or even years to conduct experiments and explore the environment around them. Outposts are small, resource-dependent, and economically unsustainable. What I mean by this is that they depend on almost all the basic necessities such as water and food to be imported, and while the discoveries or research conducted may have a significant economic impact, the outposts themselves do not have an economy or private sector. Research outposts are usually state-sponsored and funded, restricted, and do not have emergent growth like a settlement. You cannot move to the International Space Station or set up shop in an Antarctic research outpost.

A base can be considered to be the foundation of what will become a settlement. It is the 'base' of the structure that will grow and expand into something bigger. Economies are the engines of expansion. A base can have some light economic activity, or at least the potential for economic activity. An Antarctic base may conduct research as well as host tourists. It is common for old research outposts to be built onto and converted into bases. A base is still import-reliant for basically every necessity but may begin sourcing some necessities from local resources. This is typically supplemental in nature, not sustainable, but it serves as a good foundation for sustainability as it shows what is and is not possible in the given environment.

Bases are less restrictive and may allow tourists or others to visit, as mentioned before, and because of this, they may generate private funding such as Antarctic tour companies. This light economic activity generates revenue which drives further development slowly growing our base into a settlement. Economic diversification ensues. Settlements represent a significant shift from mere bases, transitioning into more self-sustained entities. They are the precursors to colonies and are marked by increasing permanence, a burgeoning population, and more diversified economic activity. Settlements are no longer just about exploration or scientific research; they are places where people live and work for extended periods, even indefinitely. Infrastructure begins to expand and evolve. Housing, sanitation, and recreation facilities start to emerge. Transportation routes and methods get established, catering to both intra-settlement mobility and connections to other bases or settlements.

Local production is ramped up during the settlement phase. The reliance on imports starts to reduce as the inhabitants harness the local resources more extensively. They might cultivate crops specially bred for the environment, or develop techniques to extract water and minerals. Moreover, the introduction of manufacturing capabilities and technologies means products can now be produced locally, further reducing the need for imports. The social structure within settlements also begins to diversify. There's a clear establishment of governance and administration. Schools may be introduced, medical facilities may be enhanced, and local regulatory bodies may be set up. This creation of social institutions indicates that families, not just individuals, are expected to live there.

Businesses see opportunities in settlements. From hospitality to entertainment, a variety of industries can establish themselves. This encourages more migration from other areas, creating a cycle of growth and development. Cultural and social events also start to shape the community.

Art, music, and other forms of expression that narrate the story of the settlers and their experiences in this new land begin to flourish. The shared experiences of the settlers foster a strong sense of community, and a unique culture and identity begin to emerge. From here colonies are the next natural evolution of settlements. They represent a full-fledged, self-sustaining community with a distinct identity and a well-established connection to their home territory.

A colony has a matured economic system, with production, trade, and commerce being major drivers. External trade can be prevalent. There may also be the establishment of a more formal governance system, often with representation or a connection to the parent country or entity. Colonies can often influence or shape the policies of their home country, especially in areas concerning trade, defense, and diplomacy. Over time, they may develop a significant population, with generations born and brought up in the colony, holding dual identities – one connecting them to the colony and the other to their ancestral or parent country.

History shows that eventually, they tend to declare independence, and the Moon is a harsh mistress.

STAINLESS STEEL SEEDS

In the deep well of the cosmos, the Moon's stark, spectral face sits in eternal silence. A tapestry of stars surrounds it, their twinkling light like ethereal whispers against the backdrop of the universe's profound quiet. Tranquil, devoid of motion and sound, the lunar plains stretch into the horizon, bearing the scars of countless celestial collisions from epochs long past. This is a stillness older than humanity, a peace as ancient as time itself.

Then, breaking the solitude, a Lunarship pierces the lunar vista. Its engine, a sapphire flame of technological defiance against the raw forces of nature, interrupts the lunar tranquility, roaring in silent testimony of man's tenacity. A blue-green plume of high-energy exhaust erupts from the vacuum-optimized rocket engine, violently stirring the lunar dust from its four-and-a-half billion-year slumber. Then just as abruptly as it appeared the engine cuts off. Silence ensues as the towering stainless steel structure gently falls a few more inches, drifting down onto the now scorched surface.

An astronaut disembarks, his mirrored visor reflecting the surreal landscape. An emissary of Earth. A boot is planted firmly into the powdery lunar regolith– an impression that will last for more than a million years. Around him, ancient dust swirls– caught in a ballet of cosmic chaos, suspended in an airless sunbeam– slowly falling back to the surface. It will not rest for another billion years. Other astronauts follow, disembarking one by one. Here life from Earth stakes a claim, here a seed is planted, one that will grow, hopefully, into a future.

Unseen by human eyes, on the other side of the lander, large panels open to reveal a deflated habitat, the tightly folded promise of a future home. The parcel is lowered and the crew of the Lunarship sets into action, delicately unfurling the compact habitat into a skeletal dome, bristling with support struts and airlocks. Powerful compressors begin to breathe life into it, filling the void with air, causing it to expand like a living organism, its walls stretching tight.

The exterior of the now-inflated habitat is studded with large pockets. Much like the pioneers of old, these astronauts make use of local resources. Using rovers, they gather lunar regolith and start to fill the pockets. This indigenous soil, once seen as a nuisance for its propensity to cling to everything, is now a lifeline. Each pocket filled acts as a shield, buffering the inhabitants from harmful cosmic radiation and micrometeoroids, and insulating against the extreme swings of temperature. As the distant sun bathes this first lunar outpost in a pallid light, the crew of astronauts set about their work unloading the Lunarship bay with a sense of quiet urgency. Even in the alien environs of the Moon, there's an air of familiarity, a choreographed dance borne of countless hours of practice back on Earth. First, the Life Support Systems. These compact yet sophisticated devices, responsible for creating an Earth-like oasis in the barren lunar desert, are carefully transported into the habitat. Next come the solar panels and batteries, lifelines to the Sun's energy. The astronauts work in pairs, painstakingly extracting the lightweight, yet cumbersome devices. Once in place, these panels will unfurl like silicon petals, eager to bask in the unfiltered sunlight, while the batteries stand by, ready to store the surplus energy for the lunar night. Communication Equipment and Computing Devices follow. Antennas are erected outside the habitat. Inside, servers and laptops are set up, their soft hum a comforting promise of an ever-present connection to home.

As the astronaut crew proceeds with the unloading, a myriad of equipment follows – food and water supplies, medical equipment, scientific instruments, exercise apparatus, and more. Each item is carefully transported in the reduced lunar gravity and meticulously placed in its rightful position within the habitat. A crew member unfolds a table upon which maintenance tools are arranged in neat order, their metallic surfaces catching the artificial light inside the habitat. Critical radiation detectors are placed strategically, their silent vigilance a shield against an invisible but ever-present danger. Nearby, EVA suits are stowed, a reminder of the harsh world just beyond the habitat's walls. Personal items, tokens of Earth, are brought in last. Photographs, books, trinkets, odds and ends, each a poignant link to home, adding touches of warmth and humanity to the otherwise functionally optimized yet bleak lunar outpost.

In this symphony of activity, the lunar base transforms from an inflated shell to a fully equipped, habitable abode, a testament to human ingenuity and ambition. Under the unblinking gaze of a billion stars, humanity becomes a permanent resident of the cosmos. The crew returns to the ship to find it empty, hollowed out. Every amenity removed, transferred to the inflated base. The metallic vessel, now a husk. They look around in quiet reflection until one of them finally says:

"Damn… why didn't we just stay on the ship?"

Everything a lunar outpost would need Lunarship already has. Instead of spending billions of dollars and years researching and designing a transportable lunar habitat just send a Lunarship to the Moon ahead of the astronaut's arrival and voila: you have a structure cheaper, bigger, and stronger than any expanding prefab you might carry with you.

Whatever structure you use has to be buried under three meters (~10 feet) of lunar regolith to protect its inhabitants from cosmic rays. This is a lot of material, heavy even in the low-gravity environment of the Moon. To support this amount of weight any structure will need to be made of aluminum or steel. Lunar temperature fluctuations warp aluminum. Steel it is. We'll expand on these topics later but for now, just understand that a retrofitted Lunarship will be the structure from which our outpost is made. A stainless steel stellar seed.

The colonization of the Americas is not a great historical reference for lunar development because Europeans could hop off their ships and immediately start using in-situ resources, chopping trees to build palisades and beds, tables, walls, and roofs. They did not need to carry their home with them on a ship to the new world as they could create amenities right there. Not to mention all the other things like the relatively stable temperatures, the relative abundance of food and fresh water, the lack of radiation, and the somewhat important ability to breathe. A better historical analog to lunar development is Antarctic development, but even Antarctica has a breathable atmosphere, abundant water, and tasty penguins. It is after all on Earth. The Moon is not. You have to carry your house with you and you can't make anything without advanced technology. Imagine if the settlers needed special 'new world' saws just to cut the logs from which they built their cabins, not to mention all the other "special stuff" astronauts need.

Despite being an inaccurate model full of historical baggage (to put it lightly) there still are some useful lessons and parallels that can be drawn from this unique period such as the importance of ships serving as early shelters during the construction of settlements. Jamestown, established in 1607, was the first permanent English settlement in America and used ships both for transportation and as temporary shelters. When the settlers first arrived, they moored their ships and lived aboard them until more permanent structures could be built on land. In only one case, however, was a ship turned directly into a structure which just so happens to be the first European structure in the Americas. After Christopher Columbus's flagship, the Santa María, ran aground on the coast of present-day Haiti on Christmas Day in 1492, the decision was made to use its timbers to construct a fort and settlement named La Navidad. Columbus left behind 39 men at this fort while he returned to Spain on the Niña. When he returned on his second voyage, he found the fort had been destroyed and none of his men were left alive after fighting with the local Taíno tribe.

Anyway, the point is that Lunarship will be used for the first two steps of development, both the survey missions and as the first outpost. For the outpost, however, it's not as simple as just dropping a ship on the Moon and calling it an outpost like beaching a ship and calling it a fort. We have to turn it sideways and bury it under at least three meters (about ten feet) of lunar regolith if we want to stay in it longer than three weeks. Why? Radiation. Earth's magnetic field and atmosphere shield it from dangerous radiation and safeguard life on the planet. Galactic cosmic rays from distant exploding stars are a significant concern, coupled with particles created in the lunar soil due to interactions between these cosmic rays and solar energetic particles.

Even more concerning are coronal mass ejections, also known as solar flares. A notable instance in human space exploration occurred in August 1972 when a series of potent solar flares burst for over a week. One such flare, an X-class from sunspot MR 11976, was so powerful it disrupted Earth's energy and communication systems in parts of North America. Thankfully, Apollo 16 had already returned, and Apollo 17 hadn't departed yet, preventing potential astronaut fatalities from this intense radiation. But it was a close call and is now referenced as a worst-case scenario as even an hour of exposure without adequate protection during an event like this could be lethal. According to NASA, the standard radiation dose for a person on Earth is about 0.36 rad a year. The Apollo astronauts received (an average) radiation dose on the skin of 0.38 rads (equivalent to two head CT scans) over the course of just 12 days.

In 2020 Chinese lunar lander Chang'E 4 recorded the first-ever detailed measurements of radiation levels on the Moon and found that astronauts in a spacesuit would be exposed to around 60 microsieverts of radiation every hour, roughly 200 times what is experienced on Earth. Thus the Lunarship hull itself is not enough to shield a permanent base against the harsh cosmic rays and temperature fluctuations of the lunar environment. Solution: bury your house under dirt to keep it cool and safe; specifically three meters of regolith as this amount has been shown to have an acceptable radiation reduction factor. This also protects against potential micrometeoroid impacts, although these would be very rare (10−6).

To do this would require moving roughly 5000 cubic meters of regolith around the Lunarship. It takes about five minutes to shovel one cubic meter of Earth in stable soil, thus shield construction would take twelve astronauts about 35 hours to complete. Not terrible, not great. Practically though we need to go all-in and send up a remote-controlled excavating rover fleet as it's going to be hard to develop the Moon by hand, and we're just getting started. This is the first big step after developing a Lunarship and performing a few survey missions.

Further radiation protection can be increased by storing all water in the base around the inside of the hull and adding an additional few centimeters thick layer of Polyethylene (plastic) which has been shown to be an efficient radiation shield.

As far as landing a Lunarship on its side is concerned, the fact that Starship is designed to perform a belly flop maneuver during Earth re-entry does mean it has the capability to handle aerodynamic loads in that orientation. The structures and tanks are designed to accommodate the stresses and strains associated with this maneuver. Also, the Moon's gravity is only about 1/6th of Earth's, which means the forces acting upon the ship during descent are substantially reduced. And of course, there's no atmosphere on the Moon, which means there aren't aerodynamic forces to contend with during landing. This makes controlling the spacecraft's descent much simpler in some ways, but it also means no aerobraking or aerodynamic control surfaces can be used. Shock-absorbing legs or bumpers would need to be added for horizontal landing. These will have to extend far enough to protect the spacecraft's body and be strong enough to absorb the landing forces without causing damage. A single-use crush canister shock absorption system is probably best as hydraulic systems are more prone to failure. Also, the interior will need to be renovated for horizontal dwelling while still keeping the mass balanced during vertical takeoff. Since the ship will be taking a one-way trip it won't need as much fuel and its engines can be retrieved and taken back to Earth later for even further reduced cost.

Finally, our outpost is going to need radiators. In 2023, the Indian Moon Mission's ChaSTE (Chandra's Surface Thermophysical Experiment) probe revealed an astonishing discovery about the Moon's temperature gradient. While investigating the thermal profile of the lunar topsoil around the pole, the probe discovered a rapid and dramatic temperature drop just beneath the surface. In particular, the surface temperature dropped from +50°C to a chilly -10°C within just ten centimeters of depth. Even more astonishingly, a decline from +20°C to -10°C was observed in a mere 60 millimeters. This discovery suggests that the lunar regolith is an incredibly strong insulator, effectively containing heat within an extremely thin surface layer, meaning it does not conduct heat well at all.

Such an unexpected finding holds significant implications for potential lunar outposts and structures. The efficient insulation of the Moon's surface layer implies that any heat generated inside a lunar outpost, base, or any structure– whether from human activity or equipment operation, may not be sufficiently dissipated away from the structure via conduction. This necessitates the need for thermoregulation strategies, particularly the use of radiators.

Heat dissipation in space is extremely important and often

overlooked. On Earth, our atmosphere conducts and convects heat away from everything. In a vacuum, a system can only lose energy through radiation. Even one lightbulb, given enough time, can generate enough heat to significantly raise the temperature in a small, enclosed space. Without proper thermal dissipation, the heat generated by an astronaut's body can continuously accumulate to the point where the astronaut can literally be cooked alive by their own body heat.

To solve this during the Apollo missions astronauts donned Liquid Cooling and Ventilation Garments (LCVGs), which were integral to their spacesuits. These garments contained a network of tubes filled with water, which circulated around the body, absorbing excess heat. The collected heat was then expelled into space using a sublimator—a device that took advantage of the vacuum of space to turn the heat-carrying water into vapor, effectively cooling the circulating fluid. The International Space Station (ISS) employs extensive radiator systems to control its internal temperatures. Large radiators that might be mistaken as solar panels expel the unwanted heat, generated by both the sun's rays and onboard equipment. These radiators are filled with liquid ammonia that circulates, picking up excess heat and radiating it back into space. This also means this technology is well established and understood so it's not really a challenge to our outpost aspirations, just an additional step that must be taken.

Alright, outpost establishment preliminaries out of the way, without further ado, dear reader I present: More preliminaries. In order to establish an outpost according to the method soon to be laid out, six things need to happen. First, as mentioned before Starship, or a Starship-like launch vehicle, must exist as a mature technology. Lunarships.

Second, generation IV Triso-fueled micro nuclear reactors must also exist as mature tech able to fit into a standard 40 foot shipping container. Through the efforts of Xenergy, BWXT, Westinghouse nuclear, and others, especially those awarded funding under the US Department of Defense's 'Project Pele,' this tech is also on track for 2030 (more on this in the next chapter). Third, research, design, and testing of a lunar environment excavating vehicle, such as a remote controlled backhoe loader. Fourth, also mentioned before, survey missions need to happen, something like Artemis. Man has not stepped foot on the Moon in over 51 years. We should not immediately jump to base creation. Walk before Running. Fifth, the belly flop sideways landing maneuver needs to be successfully demonstrated on the Lunar surface before being attempted with precious cargo. Again, the fact that Starship is designed to belly flop during its Earth descent stage, where the gravity and turbulent forces of the atmosphere are much stronger than on the Moon bodes well for the viability of this maneuver. And, finally, the outpost itself needs to be designed from a Lunarship capable of performing one of these belly flop landing maneuvers. Unmanned, one arrives ahead of any astronauts. It is retrofitted with side thrusters and extending landing legs (or you can just use a crane if retrofitting proves to be impractical). Here it will stay. It is no longer a vehicle, but a vessel, a ship turned structure. This will be the outpost. Once the dust has settled another unmanned ship arrives. Its payload is so large it must also land sideways. Inside is a lunar excavation vehicle and a lunar lorry, but most importantly it carries a 2-megawatt micro nuclear reactor housed in a 40 foot shipping container. Like the outpost before it, this ship too will never leave.

Of course it would be cheaper if this ship were reusable, able to unload its cargo and then return to Earth, however due to the geometry of its payloads it must land sideways and open like a cargo plane. Even though launching from this position then orienting vertically is technically possible it would increase risk, complexity, and require an increased fuel allowance, decreasing the payload weight capacity such that a second cargo ship launch and return would be necessary, greatly reducing the costs saved through reusability. As if that were not enough, the nuclear reactor- despite being housed in a 40ft shipping container, does still need to be housed in an exterior shelter to protect it from the harsh lunar environment and allow for maintenance access. So, like the ship-turned outpost, this ship is not wasted in a one way trip, as it maintains reusability through its use as a now shelter for the nuclear reactor that will power not only this initial outpost but all future expansions as well.

Both the lunar lorry and the excavator are primarily remote controlled, but the lorry also has a cabin allowing it to function as a transport vehicle for humans. A third ship arrives, landing normally. Vertically. This one is a human lander, carrying the lunar engineers tasked to construct this outpost. Most of their work can be accomplished from the comfort of their ship, controlling the vehicles remotely. This ship effectively acts as a construction headquarters. But here they are not safe. The ship's steel hull is not thick enough to protect them from the ambient radiation. It is safe for no more than 14 days, therefore their mission will last only two weeks. Two weeks to turn these ships into a permanent outpost. The engineers get to work.

The first task is to remove the nose cone, revealing an airlock. Inside, the now empty fuel tanks have to be off-gassed and cut open to make room for habitation. Panels stowed in the fairing are laid out like floorboards, creating a flat surface to stand on. Outside, the ship's engines needed to be removed and transported into the human lander which will return them to Earth for reuse, greatly reducing costs. A radiator array is erected and integrated. To shield the outpost from radiation, making it habitable for more than two weeks, it will need to be buried under roughly 14,700 cubic meters of regolith. This is by far the largest task. A small backhoe can excavate and transport 50 cubic meters of regolith every hour. It needs to move 14,700 cubic meters. This will take 12 days, the majority of the two week mission duration.

Once this is accomplished, final inspections are undertaken. The engineers then depart. The outpost will remain unoccupied for more than two months where it will be constantly monitored during multiple lunar day/night cycles to see how the structural integrity holds up to the drastically fluctuating temperatures between the freezing cold night and scorching hot day. If there is a failure or breach in the hull it is best that it happens before humans are present. It may not look like much now, but it's a start. A life-bearing stainless steel stellar seed, planted in barren lunar soil. Slowly over time it will grow… into a bright future.

When at last the sun breaks over the tranquil lunar horizon, man will return. This time to stay.

I *TRISO* HARD BUT IN THE END

It doesn't even matter that space is the next logical frontier and the Moon is perfectly positioned to be a stepping stone for humanity to make life multi-planetary, mine the asteroid belt, and spread across the solar system if we can't generate electricity once we're there. On the Moon, there is no wood. No coal. No oil. No natural gas. No wind. No rivers. Nothing. But there is an abundance, an excess, a plethora, an absolutely overwhelming amount of raw unfiltered sunlight– free energy blasted across the lunar surface cooking anything and everything during the 14.75 earth-days-long lunar day; immediately followed by 14.75 earth-days of freezing cold energy-sucking darkness during the lunar night. Thus we need a way to keep the lights on during the night so our people don't freeze and suffocate to death resulting in more than a few negative Google reviews.

But the constant light beaming across the slim tops of crater ridges offers a solution, constant energy 24/7- or uh, I guess 708/29.5 on the Moon. Since this electricity is intended to power life support systems and stuff, a better way to think about this is in terms of how much habitable space we get per watt. The International Space Station has an internal pressurized volume of 916 cubic meters (equal to that of a Boeing 747) hosts 7 astronauts continuously and uses 80 kw of power to cool the station and power all the life support and communication systems. In other words it costs 87 watts to keep a single cubic meter habitable, and each human requires about 100 cubic meters. Let's just round up to 100 watts per cubic meter. So for a 1,000 cubic meter habitat- a little larger than the international space station, as large as starships fairing, we'd require 100 kilowatts of power continuously.

But the problem with the crater rim approach is this solution is limited to a certain predefined geographic area. First we need to understand the scale and scope of what we're actually talking about. Let's use Shackleton crater as our case study. The entire crater isn't constantly lit, the sun creeps along the horizon and the light essentially crawls around the crater lighting up a little less than half at a time. Of course you might think "well, just put solar panels all around it" but Shackleton crater is huge, it's 21 kilometers across, about the size of a city, and deeper than the grand canyon so it's not an easy task to just cover the entire thing in solar panels, and even if you did only about half of them would be receiving light at any given moment. But there are a few unique special spots around the crater that do receive an abnormally high amount of sunlight year round. Not quite endless sunshine, but almost. These spots represent the best case scenarios, in which a solar panel array elevated 10 meters (32 feet) above ground would receive only about 3 days or 72 hours of continuous darkness. This is much better than the typical lunar night but it's still much longer than our typical nights here on Earth.

The total solar irradiance is 1,360 watts per square meter, so a 30% efficient panel would generate about 400 watts per square meter. How much energy could be generated in these spots really depends on how large you can make your solar panel array suspended 10 meters high, but no matter how much you can generate your need to store that amount of energy for 72 hours at a time- remains the same. In other words storage scales with production- well technically with consumption but consumption increase is implied with generation. A 1,000 cubic meter habitat would be large enough for 7 to 10 people and referring to our 100 watts per cubic meter figure from before, would require 100 kilowatts to keep habitable. 100 kilowatts would be generated by 250 square meters of solar panels and this configuration would require 7.2 megawatt hours of storage to make it through that pesky 72 hours of darkness. For 10,000 cubic meters capable of hosting 100 humans we would need a whopping 72 megawatt hours of grid storage. For 1000 people we'd need an insane 720 megawatt hours of storage to last the shortest lunar night possible. If we import batteries the cost will be very high. While not necessarily optimized for space, just to get an idea the Tesla Megapack uses very high quality lithium ion batteries and has a storage capacity of 3.9 MWh's so we'd need about 2 of them to keep 10 people alive for 72 hours. Two batteries would cost about $2 million, but this cost pales in comparison to the ultimate cost of transporting them from Earth to the Moon. The Megapack weighs 38,100 Kg's giving us an energy density of about 105w/kilogram, and we need 2 units for a total of 76,200 kilograms. Once again referencing our $500 cost per kilogram to the lunar surface we would have a battery transport cost of $38.1 million. So about $40 million to keep 10 people alive, or $400/watt. The largest cost here comes from the fact that batteries are heavy piles of materials. But maybe we can produce batteries on the Moon domestically, to save some of that launch cost. The Moon has basically no carbon or lithium, but it does have sodium, so let's see what it would take to make sodium batteries on the Moon and if that's something we can do.

The carbon anode makes up about 20% of the battery by weight, so a 200 ton ship importing graphite would allow for the creation of 1 million kilograms of batteries, which at 100wh/kilogram would be 100 megawatts, or about 0.1gigawatts per ship import. Sodium is present on the Moon at 3,000 ppm on average, and I couldn't find an exact component breakdown by mass for sodium batteries likely because they are less popular than lithium ion but given that they function roughly similar to lithium ion, of which lithium makes up 7% of the entire battery by weight it's relatively safe to say sodium would be roughly proportional.

We can also calculate the mass percentage of sodium in the anode of a particular sodium battery chemistry. Let's go with $Na_{0.67}Fe_{0.5}Mn_{0.5}O_2$ because this sodium-ion battery chemistry is made of relatively abundant lunar resources and can be doped with aluminum and titanium to improve stability, rate capability, and cycling performance (ACS Energy Lett. 2023), both of which are also abundant on the Moon.

First we need to calculate the molar mass of the compound:

Molar mass of Na: 22.99 g/mol
Molar mass of Fe: 55.85 g/mol
Molar mass of Mn: 54.94 g/mol
Molar mass of O: 16.00 g/mol

Na: 22.99 g/mol x 0.67 = 15.40
Fe: 55.85 g/mol x 0.5 = 27.925
Mn: 54.94 g/mol x0.5 = 27.47
O: 16.00 g/mol x2 = 32

Molar Mass of $Na_{0.67}Fe_{0.5}Mn_{0.5}O_2$ = 102.795

Mass percentage of Na = (mass of Na/102.795) x 100

15.40/102.795 = 0.1498 = ~14.9%

Given that the anode likely makes up about 50% of the battery, half of 14 is 7 so that's in line with the lithium estimate, but let's say our sodium batteries would have 10% sodium just to be generous. Sodium ion batteries have power densities of about 100 watt-h/kilogram; so to make enough sodium batteries to achieve 7.2MWh's of storage we'd need 7.2 tons of sodium. At 3,000 ppm we'd need to excavate 2,400 tons of regolith to gather enough sodium for a small 10 person base. For this extraction you'll need to spend energy to excavate and melt the material to separate it and while we don't know how much energy would be involved in excavating we do know melting regolith requires about

0.5MWh/t so 2,400 tons would require about 1,200 MWhs to get capacity to store 7.2 MWHs. So the difference between how much energy you need to store and how much energy it would cost to *make* the ability to store that energy is 3 orders of magnitude. Don't get me wrong the whole point is to get to the point where we are excavating square km's worth of lunar regolith and making sodium batteries to power things, and we would use the other stuff that isn't sodium, all the iron and oxygen and silicon and everything else, it's not just wasted; but spending 1.2 GWhs for 10 humans is much worse than just importing lithium ion batteries so domestic battery production is a non-starter until we've already got a massive kilometer-consuming industry underway.

But what about fuel cells? Fuel cells work like batteries but use electrolysis to store energy in the form of hydrogen and oxygen gas making them much lighter than batteries. During the day excess solar power would be used to electrolyse water into hydrogen and oxygen gas which would need to be liquified and stored in a pressurized tank. During the night the hydrogen gas and oxygen would be recombined to create water giving off heat and generating electricity. On Earth fuel cells are more complicated and expensive to manufacture than batteries as which is why they are not used more commonly, but those that are are usually air-breathing, only storing hydrogen gas and combining it with oxygen from the air to generate electricity, heat, and water which is then expelled as waste. Additionally the hydrogen gas itself is recharged using a dewar pressure vessel, not made by the fuel cell itself, the gas is made in a facility and shipped to the cell so it's more like a fuel consuming generator, an environmentally friendly replacement for diesel generators, than a battery. But if all the hydrogen and oxygen is recaptured and fed back into the same system, so there is no loss and the golden rule of chemistry applies, then it is a regenerative fuel cell (RFC). Regenerative fuel cells are self contained systems that only consume electricity, and only release heat and electricity, just like a battery. But these kinds of systems are even more complicated to produce, more expensive to manufacture and so they haven't really seen many uses here on Earth outside of the laboratory. But they have higher specific energy densities than batteries, more watts per kilogram, and because transport costs rather than manufacturing costs make up the bulk of our expense in space the equation flips and all of a sudden RFCs look much more appealing than batteries for space applications which is why NASA has been researching and developing them extensively. A 10 kilowatt RFC would weigh about 1.3 tons for a 3 day's dark situation, meaning for our 10 person 100kw base we'd need a 13 ton RFC which would result in a launch cost of just $6.5 million. While there's no unit production price we can guesstimate by looking at non-regenerating hydrogen fuel cells to get a ballpark range. A 5kw fuel cell costs $20k, $4k per kilowatt, so we can guess our 100kw RFC would cost at least $400,000 and to cover the cost of those extra components we can just round up to half a million giving us a total cost of about $7 million for 10 people, or $70 per watt, 82% cheaper than lithium ion batteries! So RFCs are better than batteries, but how do they stack up to nuclear energy?

Current space-based nuclear reactors include the MM-RTG used to power the Mars rovers Curiosity and Perseverance, the GPHS-RTG used in the Cassini, New Horizons, Galileo, and Ulysses space probes, the MHW-RTG used in LES-8&9 and Voyager 1&2, SNAP-3B, SNAP-9A, SNAP-19, and SNAP-27 used in everything from the Pioneer and Viking probes to the Apollo 12-17 mission Lunar Surface Experiment Packages and a few others designed but not launched. RTG stands for Radioisotope Thermoelectric Generators. Wasn't that a fun list to read?

Despite all the acronyms (and the fact they do use nuclear fuel), RTGs can be thought of more as nuclear batteries rather than nuclear generators. To generate electricity they use simple Seebeck generators, a type of solid-state device that converts heat flux (temperature differences) directly into electrical energy through a phenomenon called the Seebeck effect using semiconductors. So the Seebeck generators just convert the heat produced by the natural decay of the radioisotope plutonium-238 to electricity. Basically, it's just hot rock (plutonium) in a box. Box get hot. Put magic device on hot box and voila: electricity. Don't get me wrong these are interesting but they are super inefficient, with a typical maximum efficiency of around 8%. For reference, a typical modern car engine has an efficiency between 20-40%. The advantage of these Seebeck reactors is that they last a very very long time, with basically no moving mechanical parts so they are extremely simple and impressively reliable.

Unfortunately they only produce about 100 watts of electricity and weigh about 40 kgs, or 2.5 watts per kg, worse than batteries. However they don't need any energy storage-they run even in the dark so for a 100kw 10 person base you would need to import about 1000 of them, weighing about 40,000 kgs which would cost about $20 million. The MMRTGs used in the Mars rovers cost an estimated $109 million to manufacture, but if we manufactured 1,000 of them we could likely reduce the unit cost to $100,000 or less, but this would still cost $120 million for both procurement and transportation, or $1,200 dollars per watt, *3x as much as batteries!* Fortunately since 2015 NASA has been developing the Kilopower reactor which is way better than RTGs and future designs planned for Mars are projected to produce 10 kw each and weigh 1,500kgs, yielding about 6 watts per kg. For a 10 person base you'd only need 10 of these, 15,000 kgs, or $7.5 million in transport cost, and the one produced by NASA only cost $20 million to manufacture, so producing 10 of them for a 10 person base might cost something like $2 million each, although we'd likely produce much more, lowering the cost, but just to get an idea let's say $2 million to manufacture which is a high end estimate, resulting in a total cost of about $27.5 million, or only $275 dollar per watt, *31% cheaper than batteries.* And remember we didn't count the cost of solar panels needed with the battery configuration.

But it's unlikely these designs will ever be picked up and mass produced by private companies because the private sector is already working on even more promising micro-reactors. TRi-structural ISOtropic particle fuel (TRISO) fuel modular micro reactors are currently in development and making rapid progress from companies such as Xenergy, Radiant Nuclear, BWXT, Rolls-Royce and more. These are not your daddy's giant nuclear plants but are small and modular... and yes, they're safe. In fact, they're much safer than traditional power plants because of the TRISO fuel they use and their overall design. But even if they weren't safe the Moon is already radiated and lacks an atmosphere so a meltdown would be largely self-contained to the structure the reactor is in. So here's a quick analogy which I will use to describe- in very oversimplified and unscientific terms- why TRISO fuel is safe and cool. But not too cool, nor too hot. Just right, as preferred by Goldilocks and the Three Structures Of ISOtropic Particle Fuel.

In your daddy's traditional reactors we have this stuff, our fissile material, that cannot be too hot nor too cold. It needs to be kept just right to work. Let's call this material: "Goldilocks." If Goldilocks gets too cold she dies and isn't useful to us; and if she gets too hot she has a meltdown and goes on an arson-fueled killing spree, giving everyone cancer and salting the Earth wherever she be. So we need to keep her temperature just right using modulators– think of these as a blanket, not a thick comforter or a thin sheet, just a perfectly sized blanket keeping her temperature just right. When she's just right we can extract value from her– her life source, and we do this using a heat exchanger that conducts warmth from

Goldilocks into some medium like water. When this water gets hot enough it phase changes into steam, expanding and creating pressure which blows it out. We take a turbine and put it in front of this high-pressure steam blowout, turning the turbine super fast. The fast-spinning turbine is attached to a drive shaft. On the other end of this drive shaft are many coils of copper wire surrounded by magnets. This is a generator. And this is how we juice electricity from a happy Goldie. Nuclear reactors are all about maintaining this delicate Goldilocks balance– not too hot, not too cold, just right. But just in case something does go wrong and Goldie gets too hot we surround her with a very strong structure meant to confine her violent rage, entombing her in an expensive and multi-layered coffin of hardened steel and concrete. This is known as a containment building.

Now imagine if Goldilocks underwent anger management therapy or, I don't know, maybe she went on an ayahuasca retreat and faced her demons, undergoing a spiritual awakening in which she decides that when things didn't go her way instead of having a meltdown she simply takes a few deep breaths, finds her zen and calms down. *That's TRISO fuel.* It has a negative temperature reactivity coefficient which means if it gets too hot it stops fissioning and cools down. TRISO-Goldie counts to ten and doesn't react violently to conditional changes.

The fuel is designed to act as its own containment structure– being encapsulated by three layers of carbon and ceramic-based materials that prevent the release of radioactive fission products. Additionally, each of the fuel pellets, about the size of a grain of sand, is bundled together into another sphere of redundant encapsulation allowing for even more safety. These pellets effectively act as not only the containment building but also the nuclear waste disposal container when the fuel is all used up. Finally, the pebble bed design allows operators to just drop these fuel pellets into the top where they will react for a long time, get completely used up, and then just fall out the bottom. This means you can fuel it continuously– you don't have to shut it down for a month and change out fuel rods like in traditional reactors.

So what does this mean? Well it means our reactors can (1.) be safer while (2.) operating at higher temperatures which means they can (3.) be made smaller which means (4.) they can be made in a factory according to a standardized design which will (5.) give nuclear energy the potential to have much needed economies of scale which will (6.) reduce costs. Both the reactor designs and fuel pellets themselves can be mass produced like this leveraging economies of scale to reduce cost and this means rigorous and redundant quality control measures can be implemented further increasing safety. Safer. Dynamic. More effective. Smaller. Cheaper.

There are two sizes of these reactors: Small Modular Reactors (SMRs) which are 70 feet long and 16 feet in diameter and generate a whopping 80 MWe of electricity and Micro Modular Reactors (MMRs) which at the most are 40 feet long and can generate an impressive 2-7 Megawatts of electricity (let's go with 2 MWe to be safe). Both designs are modular and use standardized designs which means they can be mass-produced at a factory with quality control and their size means they can be put in a box and trucked to any site. Including the Moon. This also means that here on Earth they can be easily integrated into the existing power grid to start generating carbon-free energy as they can easily be placed at every site where a coal power plant has been shut down and directly integrated into the existing infrastructure. In other words, you don't need to build new buildings, rezone any areas, manufacture and ship new turbines and generators, or erect new power transmission infrastructure– power lines and transformer stations. It's all already there waiting to go.

So we need to answer three questions: How much do they cost, can we get them to the Moon, and when? At the time of writing there is no 'unit price' and predicting future costs of technologies not yet developed is a complex exercise that includes many uncertain parameters and functional forms. However, if someone did want to try to come up with a predictive model that accurately estimates a range of potential construction costs for SMR projects (an overly enthusiastic grad student perhaps?) then they might start by gathering as much data as exists and then apply different production theory approaches to estimate a range of potential construction costs. They might then apply a Monte Carlo method to benchmark the cost projections assumed by the manufacturers– varying the investment costs, the weighted average cost of capital, the capacity factor, and the wholesale electricity price in simulations of the net present value and the levelized cost of electricity. Once all this is done and the results are in they should sit down and realize how much time they just wasted by failing to do a simple literature review as the first step, for if they had they would've found all that work had already been done by Steigerwald *et al.*, 2023

Among the 19 reactors analyzed by Steigerwald und freunden, two are high-temperature gas-cooled SMRs that utilize TRISO fuel in a pebble-bed configuration, while another is a TRISO fuel block microreactor. Their cost estimates for these reactors are $5,400,000, $1,550,000, and $5,771,429 respectively. Although a small sample, the average price comes out to $4,240,476. However, considering the uncertainties and the unique challenges of lunar deployment, it's prudent to err on the side of caution, therefore let's do something crazy like double the estimated cost, leading to a price tag of $10 million for our reactor.

Currently, the only potential Lunarship we have at the time of writing in 2023 (that Xlephant in the room) has a 59-foot long and 29.5-foot wide fairing which is just barely too small to fit an SMR but offers more than enough volume for an MMR which cannot be bigger than 40 feet to fit into a 40-foot shipping container (per Project Pele). As mentioned in the previous chapter the optimal way to deliver this is aboard the ship that landed sideways to unload the vehicles.

Since it landed sideways the ship is basically stuck there so we might as well use it as our nuclear reactor housing. But since the ship also delivered our excavator and lunar lorry the cost to deliver all three would be $130 million / 3 or $43 million each, however to compare it fairly with solar we'll negate using this method and just factor in a transport cost of $100 million for 2-7 MWe. Let's say 2 MWe for conservative margins which is about $55/watt, *85% less than the cost of batteries and 22% cheaper than the RFC!*

Finally, from what public information we have available these reactors appear to be on course to hit the market by 2027. Or 2028. Or 2030. Or 2032. Who knows, these are being developed by government-funded private-public partnerships so cost overruns and delays are to be expected but it's probably safe to say these will be available by the mid-2030s.

So now we have our lunar outpost, but how do we get to the next step: a lunar base? If we're talking about developing the Moon rather than just visiting then we're gonna need more room and Lunarship lacks space, specifically of the 'floor' variety. We can theoretically make our Lunarships bigger, but because they have to be aerodynamic to punch through Earth's thick wet dripping atmosphere they'll always be, how should I put this? Constrained in the girthy direction. Sure we can make the ships wider and taller but that means making the boosters wider and taller with more engines and more risk. It also means making the launch facility wider and taller and entails increasing material strength making our fairing walls thicker and heavier thus requiring even more engines, fuel, volume, and money. See, rocket science isn't hard just because it involves complex formulas. Rocket science is hard because it is a recursive balancing act involving complex manufacturing constraints and supply chain logistics. Increasing the volume of a rocket has many tons of benefits (punnage intended) but these benefits have diminishing returns when compared to the likewise substantial cost of building a structure on the Moon using resources found there, on the site: in-situ resource utilization. If we can do this we can design structures to better suit different purposes, tailored to fit certain needs just like buildings here on Earth. So, how do we start building things out of the Moon?

The answer:

CHROMIUM ANODES

Constructing a lunar base presents unique challenges, particularly given the Moon's lack of atmosphere and the resulting extreme temperature variations. During its prolonged day, the surface can reach a blistering +120°C, while at night, without the buffer of an atmosphere, temperatures plummet to a frigid -130°C. Such drastic fluctuations lead to rapid expansions and contractions of the lunar surface causing small quakes called *"thermal Moonquakes."* Four seismometers deployed during the Apollo 17 mission recorded thousands of these Moonquakes over an 8-month span from 1976 to 1977, but the poor quality of the data made analysis impossible until 2023 when a group of three scientists using modern computers developed algorithms to accurately interpret the data.

While analyzing this nearly half-century-old data they stumbled upon a mysterious, recurring vibration. It seemed almost...unnatural. The epicenter? Precisely where the Apollo 17 mission had landed decades ago. As they poured over the data, cross-referencing timestamps and analyzing waveforms, a chilling realization set in: the vibrations were emanating from the descent vehicle the astronauts had left behind. But why? What forces were at play to make an abandoned lunar module from 1972 the heart of such disturbances? Was there *something* or maybe even *someone* aboard the abandoned derelict? To this day nobody knows for sure. Unfortunately, the scientists who conducted this study (probably coerced by the deep state) tried to explain these anomalies away as *"thermal fluctuations."* As temperatures rise, the aluminum lander expands, oscillating and creating detectable vibrations on the surface, a phenomenon quite distinct from natural Moonquakes.

But others are not so sure. Independent researchers believe the origin of the mysterious vibrations might be something *otherworldly*, maybe even *ancient* and *alien*.

Whatever is causing these mysterious vibrations emanating *precisely* from the aluminum lander at the *exact* times the lander is heated and cooled by the lunar day/night cycle, the varying strength of the vibrations correlating *perfectly* with fluctuations in thermal intensity... we may never know. But one thing's for sure: future lunar bases must be constructed from materials that can withstand such thermal stresses without inducing unwanted seismic activities– alien or otherwise...

Aluminum, while lightweight, undergoes a contraction and expansion cycle that may compromise its structural integrity over time. Steel, on the other hand, with its high melting point and ability to operate at temperatures up to 820-870°C, emerges as a preferable choice. When coupled with efficient heat management systems, steel offers a reliable and resilient solution for lunar habitats, ensuring the safety and stability of its occupants in the face of extreme lunar conditions. But what is steel?

Well, it's just iron– with some carbon. The biggest difference between the swords and plows of the Iron Age and the metal skeletons holding up today's skyscrapers is how much carbon is in that iron. Cast iron contains over 2% carbon, making it extremely hard but also brittle. In contrast, wrought iron, the purest form of iron, has less than 0.08% carbon, granting it superior ductility, enabling it to flex without breaking, but also rendering it softer than both cast iron and steel. Positioned between these extremes, steel boasts a carbon content ranging from 0.2 to 2%, striking a balance between hardness and flexibility. Today, steel production is a marker of a nation's economic strength but it is also the single biggest industrial source of climate pollution. Because steelmakers still use coal-fired blast furnaces, industrial steelmaking releases about two tons of carbon dioxide emissions for every ton of steel produced–adding up to nearly 10% of Co2 emissions worldwide.

The solution: Use electricity to melt and purify the iron rather than a coal-fired furnace. If you run electricity through a cell filled with an *Electrolytic* mixture the electricity will heat the cell up to about 1,600 °C (3,000 °F) resulting in a hot *Oxide* and *Molten* iron soup. In addition to heating things up, electricity drives the *Oxygen*-removing chemical reactions. *Molten* iron gathers at the bottom of the *Electrolytic* cell, and *Oxygen* gas is emitted instead of carbon dioxide. Then they drain the high-purity *Molten* iron and add carbon as it solidifies to make steel.

This process is known as *Molten Oxide Electrolysis.* Why it's named this I can only guess but it has the potential to decarbonize the steelmaking industry. This process is currently being developed to scale by Boston Metal, a company headquartered in... yep, you guessed it: Woburn, Massachusetts. Anyway, what this has to do with developing the Moon is it potentially allows us to manufacture steel on the Moon using lunar regolith since lunar regolith consists of 5-15% iron on average. In fact, Molten Oxide Electrolysis was actually discovered in an attempt to figure out how to extract oxygen from the Moon's surface for lunar bases. See, the Moon is actually covered in oxygen, it's just trapped in the lunar regolith. Lunar regolith consists of 41-45% oxygen on average.

NASA has been funding lunar oxygen procurement for a while now and in pursuit of these fat stacks, ahem* I mean *"scientific endeavors"* many techniques have been developed to extract the oxygen from lunar regolith, but the byproducts of these techniques happened to be a garbage alloy of whatever was left of the regolith that wasn't oxygen– namely silica, alumina, calcium, magnesium, titanium, and iron which obviously got a lot of people thinking: *"hey that's quite convenient– maybe we could actually use these 'useless byproducts' to, oh I don't know... make stuff on the Moon?"*

The problem was the alumina, silica, and iron weren't neatly separated into three different byproducts, instead, it was a single clump of slag, an amalgamation of all the leftover stuff tightly bonded together, and the biggest issue was in finding a way to actually separate these leftover constituents. Enter MIT scientist Dr. Donald Sadoway who figured out the constituents could be separated by running electricity through a cell filled with lunar regolith and an Electrolytic mixture. The electricity heated the cell up to about 1,600 °C (3,000 °F) resulting in a hot Oxide and Molten regolith soup. In addition to heating things up, electricity drove the Oxygen-removing chemical reactions. Molten constituents gathered at the bottom of the Electrolytic cell, and Oxygen gas was emitted. Sound familiar? *Molten. Regolith (41-45% oxygen 5-15% iron on average). Electrolysis.*

Unfortunately, besides those fat NASA stacks, electrolyzing lunar regolith didn't seem like it was going to be very profitable (or useful) on Earth. But what about electrolyzing terrestrial Earth regolith? Interestingly terrestrial regolith (street name: dirt) consists of 0.2% to 55% iron on average and has a higher market cap derived from its locational advantage. But you know what else is abundant on Earth and has an even higher iron percentage on average than terrestrial regolith? Terrestrial Iron Ore... which is used to make steel. Incredible, I know. So Dr. Sadoway got to work developing the process that would become known as Molten Oxide Electrolysis, electrolyzing iron feedstock to make carbon-free 'green steel.'

But there was an issue. The graphite anode was consumed in the process, making the process too expensive since you had to replace the anode each time. Losing the anode with each run would be like having to replace your lightbulb every time you turn on the light for a few minutes. So Dr. Sadoway tried to find a material that wouldn't be consumed and found that iridium worked, but since iridium is extremely rare it remained too expensive for any applications outside of a research lab. Then a metallurgist, Professor Antoine Allanore, joined the team and together they found that an anode made of a chromium-iron alloy (cheap and abundant) worked efficiently without being consumed. So they went and started Boston Metals which has since scaled up this process from a small lab cell using less than 5 amps of electricity and producing just a few grams per day of molten iron to a bus-sized cell that uses 2500 amps and produces tens of kilograms per day. Currently Boston Metal is validating its full commercial design which uses 25,000 amps and produces hundreds of kilograms per day.

Now you might be thinking okay I get that Boston Metals' foundational technology was created in pursuit of supplying lunar bases with oxygen but their target market is the terrestrial steel industry right... So what does this have to do with industrializing the Moon? Well, Donald Sadoway is a founder of both Boston Metals and Houston-based company Lunar Resources which outright states their intent to industrialize the Moon. This is yet another great example of how NASA funding research in what to some may seem to be far-flung vanity projects foolishly reaching for the stars (while we have so many problems here on Earth) usually results in amazing solutions to those same terrestrial problems so often brought up as objections to the "*vanity*" funding.

So what's the difference between Molten *Regolith* Electrolysis and Molten *Oxide* Electrolysis? Well, it's *basically* the same technology and in lunar development, the terms mostly represent a 'step distinction'. When you put regolith into the cell out comes oxygen gas, aluminum, silica, and *low-purity* molten iron. Molten Regolith Electrolysis. Then you can take that low-purity molten iron and feed it back into the cell, refining it into *high-purity* iron. Molten Oxide Electrolysis. The difference is the purity of the iron, and to make steel we need very high-purity iron.

But very high-purity iron is wrought iron, not steel, so we need to add carbon to make steel. Additionally, there are different types of steel with each having different grades of that type. Traditionally different types of steel are made by *removing* carbon to a certain percentage level and then adding alloys such as nickel and chromium to yield certain types of steel such as stainless steel or carbon steel. This step in the process is usually done once the molten iron has been moved into a ladle, basically a giant 'pot' to carry all the molten liquid which is why this step is called *ladle metallurgy*. However, with the electrolysis process, we'll actually need to *add* carbon to a certain percentage level in addition to alloys as the electrolysis process doesn't use coal and the lunar regolith basically has no carbon to begin with.

The biggest limitation to full-scale lunar steel production will be the amount of carbon we have since extremely pure carbon needs to be added to the high-purity molten iron as it cools. Unfortunately, carbon is scarce on the Moon. Luckily we have approximately 1.85 quintillion tons of it here on Earth, and humans are made of so much carbon that they literally just spew it everywhere they go; every fingerprint, every breath, every bowel movement– carbon just exhaling and oozing out of every orifice. So it's not really a hard problem to solve.

The next step is to cast the steel into a shape using a mold and then machine those shapes into other more useful shapes. Also if we need to off-gas the molten steel at any point before casting then it's helpful that the Moon has enough vacuum suction power to make James Dyson blush. As for electricity, our nuclear reactor provides more than enough. If our outpost uses 100 Kw and our reactor produces 2 Mw we'll have 1.9 Mw left. The electrolysis cell that can produce hundreds of kilograms of molten iron per day uses 25k Amps. The electronegativity of iron is 1.83 and oxygen is 3.44. Since electronegativity changes with temperature (thank Josiah Gibbs for that) and we'll need an overpotential, a realistic setup would use about 1.5 to 2 volts. To be safe let's go with 4 volts which would result in a cell that produces hundreds of kilograms of steel per day requiring only about 100 Kw to operate leaving us with 1.8 Mw left!

This technology can be scaled up or down to nearly any size, but the hundreds of kilograms a day capacity requires a volume of about 730 cubic meters. Lunarship has a volume of about ~1,000 cubic meters, so it fits. Since the steel-making process involves two electrolyzing steps we should specialize and deliver two of these. For clarification these cells will be delivered by a Lunarship that will return to Earth, not one intended to be converted into a base. Since these reactors will be dealing with sharp radiated regolith it's probably best to keep them outside the shelter. You could literally just plopping them down exposed on the dusty lunar surface. They'd be fine, but it would be better to shelter them a little so the people working with them will be exposed to less radiation. Fortunately there's a bunch of unused space in the empty fuel tanks of that nuclear reactor ship so just cut them open and pop those bad boys into the back there

Once you have an electrolysis cell full of high-purity molten iron ready to go you take a drill and drill a hole into the plug holding all that liquid in. This is called 'tapping.' Once the cell is tapped and metal is flowing it needs something to flow into which, as discussed before, is where the ladle comes in. This will be about the size of a 100-gallon water trough but with much thicker walls, and it should be able to be lifted by two or three astronauts while full. Keep in mind on the Moon things weigh ⅙ what they do on Earth. 100 kilograms would weigh 16.5 kilogram. Once you've got this pot o' iron you sprinkle in carbon and stir. (Optional: add a dash of alloys to make different types of Steel.) Now your pot o' iron is a pot o' steel and we need to pour it into fun and useful shapes. We do this with a caster. This is basically a space lined with rollers that are water-cooled. When we pour our liquid steel into this space it comes into contact with the rollers which exchange the heat in the steel, through the rollers, and into the water which flows away. The energy in this water can be used for something else or just fed into a radiator. When the steel is cooled enough (but still *really hot*) it becomes malleable, almost clay-like, and rollers massage it into a long square or rectangular beam.

In industrial steel production plants here on Earth steel beams come out of the caster still *really hot* so they can be turned into even more shapes later on and then cooled off. But we're not talking about an industrial steel production plant capable of making millions of tons of steel of a thousand different varieties and grades into hundreds of different shapes annually. I mean, we're on the freakin' Moon so let's just try to make some simple beams and panels to start. Panels and beams make dreams! The caster will output beams but for panels, we'll need a press.

Then we're gonna need somewhere for our molded steel beams and panels to roll onto when leaving the caster so we'll need to add a table with passive rollers to our order. This can be foldable and doesn't need to be powered, so not a big deal just something to note. Once a steel beam is produced from the caster and squeezed out onto the roller table it will need to be taken *somewhere*. I don't recommend just dropping these steel beams into the lunar regolith or storing them aboard our already cramped quarters so let's just add a cantilever rack or two to our list. So we have a few ladles, a caster, a press, a roller table, and a few cantilever shelves which should fit into the ~200 cubic meters of fairing volume we had leftover on each of our electrolysis cell trips (two trips for two cells with an excess space of ~ 200 cubic meters = ~400 cubic meters left over to transport this 'stuff'). The largest items will be the casters since they're long but they can be broken up into pieces and assembled onsite. Now we have the capital to make steel.

The final step is to set up one more premade Lunarship structure that comes with a 5-axis CNC machine, some 3D printers, and some work surfaces built into a horizontal deck according to a premade floor plan. Drop, flip, and bury it next to the first Lunarship, and voila: a machine shop. We have just unlocked the ability to produce *things* out of *dust*.

This is a huge next step just kinda crammed into one paragraph at the end of a chapter so let me clarify: we have one Lunarship buried and turned into an outpost. This outpost has been receiving supplies from other Lunarships that return to Earth (and actually function as 'ships'). So far we have a nuclear generator and two steel-making electrolysis cells. Then, in this last step we sent another Lunarship to be turned into a structure adding a second 'building' to our outpost. This 'building' is a machine shop/workshop. Now that we have steel we can machine it into tools and stuff. Amenities. Things. Struts, nuts, bolts. Whatever. Now that we can make steel beams and steel things we can build our base *out of* the Moon. So how do we do that?

Well, as with any construction project, the first step is to...

BREAK GROUND

And don't stop until you've dug a big square hole. How big? Big enough to give us more space of the floor variety than a Lunarship has to offer. I recommend going 20 meters by 20 meters, and four meters deep. Hopefully, you sent a remote-controlled excavating rover fleet the first time you had to make those regolith radiation shield walls. The reason we're digging so wide and deep is because we're going to put a *big* building in this hole. This way the walls of the hole will be the lateral radiation shield, except instead of three meters thick they'll be a mile thick. Now you might be thinking digging a hole is more work than just scraping and piling regolith around the structure like we did for the two Lunarships we turned into structures but digging a hole actually requires less work once we get to a certain building size. A 20x20x4 meter cubic *hole* would require displacing 1,600 cubic meters of regolith. A 20x20x4 meter cubic *surface structure* would require gathering 2,304 cubic meters of regolith to pile around the structure to achieve the necessary shield thickness. Digging a hole is equivalent to gathering the roof/top shield material (plus a meter) but comes with four "free" walls. While digging we may run into boulders and other complications however the same is true for scraping roughly 2 Kilometers of the Moon's surface which, by the way, isn't nice and flat but rather rough uneven boulder-strewn terrain.

Once this hole is made habitable by filling it with steel and polymers and gas and stuff, the regolith we displaced to dig this hole can then be dumped on top of the structure providing a shield ~4 meters thick atop the structure. This structure needs to be rudimentary. This is the first true 'construction' project off-world and unforeseen complications will arise. The techniques and lessons learned here will allow us to expand our construction knowledge and hone our capabilities for bigger more complex projects going forward, but to minimize our troubles this first structure should be as simple as possible to flatten the learning curve. Essentially it's a prototype, a demonstrator. So what's the simplest way to turn a hole into a home? Well, thanks to our electrolysis cells we have steel beams, panels, and oxygen. But pretty much everything besides that will need to be imported from Earth. Fortunately, the heaviest part of any building is the concrete foundation and steel support, both of which we can make on-site.

Lunar concrete, or *lunarcrete* as it's come to be called, is probably unjustifiably costly, at least for *this* project. The basic ingredients for lunarcrete would be the same as those for terrestrial concrete: aggregate, water, and cement. In the case of lunarcrete, the aggregate would be lunar regolith. The cement could be manufactured from the silica, alumina, calcium, and magnesium 'slag' content left over from our Lunar Regolith Electrolysis step. But the fatal flaw of lunar concrete is the need for either lots of water or lots of sulfur, neither of which is particularly abundant on the Moon. Water is almost certainly out of the question as it is too precious to waste on concrete production; however, sulfur can be used in its stead and there is natural sulfur on the Moon. In fact, the 2023 Indian Moon Mission's six-wheeled Pragyan rover detected significant concentrations of sulfur in the polar regolith, but *"significant"* in this case is still a tiny percentage of lunar regolith by weight. How much? Well at the time of writing, we still don't really know; which means it's hard to calculate whether or not we can realistically gather enough of the stuff to justify the cost in equipment and space that would be required to make lunarcrete.

Fortunately, we don't necessarily need a lunarcrete foundation as our structure will have a wide base made of steel and be held down by the tons of regolith piled atop it. Basically, it boils down to *if we can* easily gather lots of sulfur then great let's make it. Perhaps our Lunar Regolith

Electrolysis step will yield lots of sulfur-infused slag. But *if we can't* easily gather lots of sulfur to make lunarcrete then it's not the end of the world. Also, it's worth noting that any sulfur we do gather may have higher priority uses as it can be used to make sealants and fertilizers. Anyway going forward let's assume we can't make lunarcrete so the base of our structure will literally just sit on the surface of the hole. If we feel we *need* to anchor it in place for whatever reason we can drive steel beams deeper into the ground at several points, staking it in place.

Alright so we have a gaping hole that needs to be filled, so the first step is to create a foundation by laying steel beams in a grid with vertical beams rising up at all vertices. Or as many beams as is necessary to support the roof plus the several thousand tons of regolith that'll be piled on it. Ya know, just do a basic load-bearing analysis and then multiply it by 1/6th. Then use a hot glue gun to join all the beams together (or maybe use fasteners or cold welding?). Once dry and stable, create a mirrored grid at the top: a roof, and glue them together. Now you have a big scaffolding from which you can affix steel panels too. Just cover the whole thing in steel panels, making a big steel box. That's what buildings are right? Big steel boxes? Well at least on the Moon they are since you're going to have a damned hard time finding enough trees and plaster to do anything else. Anyway, the Lunarship structure we've been living in up to this point is basically just a big steel tube so this is pretty much the same thing. Next hermetically seal the box shut so it's airtight as air is kinda important on the Moon. Wiring, plumbing, floor material, door seals, windows, water filters, what shade of white carpet we should put in the bathroom, and where we should eat out tonight are all details I'll leave to the engineers. Again, we're focused on development, not design. But here are some notes to keep in mind:

First, store all the water in the roof and walls for the same reason we did in the Lunarship structures: to combat secondary radiation. Remember the regolith itself is slightly radiated so you don't want to be directly touching walls that directly touch it. A layer of Polyethylene a few centimeters thick acting as a bladder for water storage should do the trick. The next thing to note is we'll need to actually be able to get in and out of this steel box so don't forget to make an airlock access point somewhere at the top. Lights, life support, ventilation, ducts, copper wiring, gaskets, pipes, pumps, rails, nozzles, crown molding– most of these will have to be imported from Earth but a lot of things can be made in our fancy 3D printer-equipped Lunarship machine shop. Also, as with the Lunarship structures, we're going to need lots of radiators to keep this steel box from turning into a big oven; however, we may be able to make these mostly of steel, saving much expense. Then pressurize it with 40% nitrogen imported from Earth and 60% with the oxygen produced by the electrolysis reaction when making the very steel this box is made of. Isn't that cool? Both the building itself and the air inside it will have been made out of Moon dust. Note on nitrogen imports: The internal atmosphere could be 100% oxygen but this poses a substantial fire risk. Modern spaceflight missions use a 40/60 nitrogen/oxygen mixed-gas composition to reduce the chances of a flash fire as well as to eliminate the complications of purging nitrogen from an astronaut's body prior to launch to avoid decompression sickness (the bends). Fortunately, any imported nitrogen is nearly infinitely recyclable and very inexpensive making the cost insignificant. But what should we do with this building? What is the optimal use of 400 square meters? Well, how about a place to store food and water?

(P.s.i. If you're worried about pressure bulging because we made this structure a 'cuboid' just remember that the internal pressure will be around 14 psi. Structural steels typically have tensile strengths in the range of 50,000 to 100,000 psi. Think about cuboid fish tanks. Beveled edges and tension cables are additional solutions.)

FOOD & WATER

In 1969, when the Apollo astronauts made their historic return from the Moon, the prevailing belief was that the lunar surface was devoid of water. This perspective began to change as a series of missions over the subsequent decades started to paint a different picture. For nearly 20 years, various orbital and impactor missions undertook detailed studies of the Moon. One such mission, NASA's Lunar Crater Observation and Sensing Satellite, confirmed the presence of ice within the permanently shadowed craters at the Moon's poles. These are areas that never see sunlight, making them incredibly cold and perfect traps for water ice. Parallel to these discoveries, several other space missions began uncovering hints of hydration beyond these cold, dark regions, identifying signs of water even in regions that do receive sunlight. A pivotal moment in our understanding came in 2020. NASA's Stratospheric Observatory for Infrared Astronomy (SOFIA) made a groundbreaking confirmation: the presence of water on the sunlit areas of the Moon. This wasn't just a minor discovery; it suggested that water on the Moon wasn't limited to its cold, shadowed recesses, but could be more widespread than previously believed. By 2023, building on these revelations, SOFIA had crafted the first water map detailing the amount of water distributed across an area about one-quarter of the Moon's Earth-facing side.

So the Moon has tons of water, all we have to do is scoop it out of crater bottoms or steam it out of the regolith and we can just continue on our merry way developing the Moon, right? Of course, it's not so simple. SOFIA confirmed water even in sunlit areas to the order of about 100 to 412 parts per million: or about 12 ounces of water per cubic meter. The bottom line is that even the wettest parts of the Moon are still 100 times drier than the Sahara desert. To top it off another recent 2023 study found that heating regolith to 'steam out' the trapped water doesn't work nearly as well as previously hoped. The regolith, being such a strong insulator, basically cakes into a hot hardened layer around the outside of the heated mass impeding ambient conduction and trapping water inside this shell. This can be solved by continuously stirring or vibrating the regolith but now we're talking about even more energy input for a tiny amount of water.

But hey aren't we already heating up regolith to extreme temperatures using tons of energy to create steel in the electrolysis reactors? Precisely. Besides oxygen, the regolith electrolysis step of our steelmaking process will undoubtedly also yield trace amounts of hydrogen and water vapor which can be captured. However, because the amounts are so low this will essentially be supplemental in nature– not a source we can realistically depend upon to water our people and plants.

Likewise, lunar ice prospecting could prove useful but also dangerous and logistically challenging. Venturing into the bottom of billion-year-old craters might be just fine. Or it might prove fatal, with loose deposits of regolith and boulders undisturbed for millennia suddenly becoming dislodged and tumbling down in a Moon rock-regolith slide. So unless we can confirm the exact location of a few dozen deposits of ice at least a few tons in size this source will also be supplemental as we shouldn't rely on the chances of ice prospecting to keep our people alive. Maybe we'll find some ice while digging and moving all those tons of regolith for our Lunarship shields and holes. But if not we'll need to bring every liter of water needed to the Moon from Earth. Fortunately, we have the technology to recycle almost all water used once brought up. In 2023 the International Space Station's Environmental Control and Life Support System (ECLSS) was upgraded to achieve a record 98% water recovery rate. This system treats and recycles all water aboard the ISS including sewage, wastewater, and even humidity scrubbed from the air. The small 2% loss rate means that the two aforementioned sources of water procurement, possibly even just the water recovered from the electrolysis cells alone, should be enough to counter this loss.

We'll need to send about five liters (1.3 gallons) of water per person up from Earth for basic necessities. If we want to live more comfortably with hot showers and stuff it's more like 100 times this at around 200 liters (52 gallons) per person. An entire Lunarship full of water would provide 200,000 liters, or enough water for 40,000 people to live like they're on the ISS (no showers) or enough for 1,000 people to be able to enjoy showers and stuff like on Earth. Using our $100 million launch costs this would be $500 per liter, roughly the same price as concert venue water bottles. Put another way, 1000 people can live comfortably on the Moon for about the same price as a private jet. Every delivery increases our carrying capacity. Recall this water also plays a vital role in radiation shielding as it will be stored in the hull of our structures providing an efficient barrier against secondary radiation.

Food is harder. When it comes to farming the issue isn't whether or not we CAN grow food on the moon, it's how many calories you can yield per cubic meter. Is a cubic meter of moon-grown calories more expensive than a cubic meter of Earth-grown and rocket-transported calories?

If the answer is yes: import food from Earth.
If the answer is no: Farm on the Moon.

So first: How many calories can you yield per cubic meter? With highly efficient modern farming technology on Earth 10,000 square meters (1 hectare) can yield enough calories to feed an average of five (five!) people per year. But on the Moon, and pretty much everywhere else besides Earth, we have to do a lot more work than simple soil processing to grow crops. We have to hermetically enclose a space and then pressurize it with the right amounts of oxygen and nitrogen and carbon dioxide, add a regenerative closed-loop life support system, and then protect this enclosed space from radiation. So we need to look at hydroponic farming.

Let's consider a hydroponic farm that is 1 cubic hectare, or 1 million cubic meters. Assuming a 1 meter gap between rows we get roughly 50 hectares of food, or enough to feed 250 people a year. But the advantage that hydroponics have over regular open-air farming is one can yield 3-4 harvest seasons a year rather than just during the spring-summer cycle. Thus a cubic hydroponic hectare farm can yield enough calories to feed 750 people annually. The average cost to build a single story warehouse on Earth in Austin Texas is about $175 per square meter on the high end. Let's round up to $300 to include furnishing it into a hydroponic farm. 1 story is about 3 meters tall, but let's go with just 2 meters and say we have a 100x100x2 meter building (1 hectare^2) then on Earth the hectare sized hydroponic building would cost about $6 million. Then again assuming a 1 meter gap between shelves, a cubic hydroponic farm would be 50 of these and would cost $300 million- which is about how much a cheap skyscraper costs. But how much would it cost to build one of these on the Moon? Well that's the hard part. Assuming we're building according to the method laid out in this book in which we use chromium anode electrolysis reactors to melt regolith into steel and build big steel boxes in holes, the cost is going to be determined by the *useful lifetime* of the reactors and excavating equipment, plus wages and transport costs. For example this first 20x20x4 meter building would cost $3.125 million per cubic meter of space made habitable.

But if you use all the same equipment and just build another building exactly like this it will cut the costs in half plus operating expenses, so maybe a 40% cost reduction from $3.125 million to $1.875 million.

Then do it again, and again, each time reducing the cost by nearly half.

(In millions: $3.125; $1.875; $1.125; $0.675; $0.405, etc.) How many times can you do this? Precisely. I don't know. It's impossible to find out without rigorous testing of all equipment involved, so let's just guess. Let's say it will cost 3x as much to build per cubic meter on the Moon. That means our $300 million dollar Earth-based cubic hectare of hydroponics would cost $900 million on the Moon. This includes hermetically sealing it and insulating it against thermal fluctuations and- if humans are present burying it under regolith for radiation protection, as well as importing water and nutrient substrate and other components. That's a lot, I know, but increasing the price only works in favor of my argument so I'm trying to be as lenient as possible while still maintaining some realism. So $900 million. Let's round up to $1 billion or $1000 per cubic meter.. $1 billion to feed 750 people annually, or **$1,333,000 per person.**

How does it stack up to importing food on a Lunarship? Requiring 3,000 kilocalories (kcal) a day, to feed one person we need 1,095,000 kilocalories a year per person. There are 3,640 kcal/kilogram of white flour, 2,710 kcal/kilogram of beef, 1,300 kcal/kilogram of rice, 770 kcal/kilogram of potatoes, 3,870 kcal/kilogram of sugar, 4,020 kcal/kilogram of cheese, 7,170 kcal/kilogram of butter, 1,550 kcal/kilogram of eggs, 2,390 kcal/kilogram of chicken, and 8,840 kcal/kilogram of olive oil (the most calorie-dense food). Using these ten examples we get an average of 3,626 kcal/kilogram of human food. A single Lunarship load of human food would provide 725,200,000 kcal, enough to feed 662 people for one year. Another way to think about this is that we can "send up" 132.4 hectares worth of food per Lunarship.

Adding in the cost of food procurement, the most expensive food group in the US in 2023 is meat with a mean price of $1 per kilogram or roughly $1 per 2,500 kcal. We're not filling our Lunarship with meat but we will count it as if we were so we have a 'most expensive' estimate to be safe.

This increased margin should also account for the cost of packaging and logistics of everything since it was already factored into the meat cost. Using this highest of costs we get a price of $290,080 to fill an entire Lunarship, and combined with the launch costs we get a total cost of $100,290,080 to feed 662 people, or ~**$151,495 per person**, an order of magnitude less than the hydroponic farm! So a lunarship can feed 88.2% the amount of people, but for basically 11% the cost, and the farm would have to produce food for 10 years before breaking even with lunarship imports assuming no reduction in launch cost, but as we'll discuss in a future chapter launch cost reductions to less than $20 million are nearly guaranteed using modern technology. If this is the case it will take the farm over 50 years to break even. It is in a race against launch costs and bound to lose.

We were also very generous with build costs, but even if we get to the point where building on the Moon doesn't cost 3x as much as on Earth but is on par with it, we still have to consider market demand and opportunity cost for those habitable spaces which will undoubtedly have much more valuable use cases and people willing to pay more for said space than you'd ever make from farming. An Iphone's value isn't what it costs to make an Iphone, an Iphone's value is what people are willing to pay for it- market demand. Same with habitable space on the Moon.

Value density. Any space not on Earth requires a lot of modification to be made habitable. We have to hermetically enclose a space and then pressurize that space with the right amounts of oxygen and nitrogen, add a regenerative closed-loop life support system, and then protect this enclosed space from radiation, and control the thermal environment with radiators. This is why I prefer to think in terms of 'Habitable Space' as it helps to form a more accurate conceptual image of what we're actually doing when developing the Moon. This might seem pedantic but it's essential. Understand that we are building bunkers not levittown homes. We don't get to spread across the heavens organically like we did here on Earth. The entire history of humanities mass migrations out of Africa– onto 7 continents, across 5 oceans– was easy compared to what we have to do now. Any activity beyond Earth requires careful planning and focused intention as we painstakingly carve small, warm, gas and moisture filled pockets out of the cold entropic abyss, one little oasis at a time. This naturally makes every cubic meter of habitable space expensive. So we have to maximize the value gained per cubic meter. Only extremely high-value activities can be profitable here.

We can see a similar effect play out in the terrestrial real estate market. Real estate downtown is much more expensive than real estate in rural areas because of market competition for proximity advantages. In January 2022, the median price per square foot in Queens, New York was $540 dollars, in Manhattan it was $1,612 dollars. If you spend $16 million dollars to buy 10,000 square feet in Manhattan it doesn't make sense to use that floorspace to grow crops. Instead you'll want to maximize the value per square foot to the point that you either break even or profit from the initial investment. Maximization of value density.

Saudi Arabia specializes in crude exportation because oil naturally occurs there. Sure you can synthesize oil from coal in Germany but why would you when you can import it cheaper from Saudi Arabia? California specializes in wine exportation because of its naturally occurring grape-friendly climate. Sure you can grow grapes in a greenhouse in Canada but why would you when you can import it cheaper from California? Sure we can make Iphones in the US but why would you when you can import it cheaper from China? Sure you can grow food on the Moon… but why would you when you can import it cheaper from Earth? This is geographic specialization on a solar system scale and it leads to an interesting long term vision. In many visions of the far future Earth is portrayed as a cramped, polluted, dying world-resources stretched thin- just barely able to keep up with humanity's demands. But now think about the far future of Earth in the context of there being a market dynamic incentivising agriculture. Eventually factories and people can be moved off world, but farming will remain. I see a vision of the future in which Earth serves as the sole breadbasket of a solar system spanning civilization, a gaia world in which environmental stewardship isn't just necessary, but part of a profitable business environment, for maintaining a healthy ecology yields abundance.

Unless of course Earth was smashed by an asteroid or obliterated in a nuclear holocaust in which case you'd probably wish you had a few backup farms on the Moon or in orbit…

THE ENVELOPE'S BACK

Let's attempt the impossible and try to calculate how much this strategy would cost. The first thing we need to estimate are our research and development (R&D) costs. Up front we are not looking at the R&D costs of developing a Starship/Lunarship type launch vehicle, these are separate because- while being able to develop the Moon does depend on this technology existing, this technology isn't being developed with the sole intent of developing the Moon. Starship's use as a vehicle for building out and maintaining the Starlink constellation alone is justification enough for its development, not even counting NASA and other government or military contract missions or the entire private market earning potential or any of our lunar development use cases. Caravels weren't developed to cross the Atlantic, they were developed to sail down the coast of Africa, but once they existed they made crossing the Atlantic possible. Same with Starship-like launch vehicles. So then what exactly are we researching and developing?

Two vehicles, the process of burying a Lunarship under three meters of regolith using one of these vehicles, lower gravity adjustments to the steel-making electrolysis cell, vacuum-steel casting, and biggest of them all: The lunar structure construction methods. Building big steel boxes in big holes on the Moon may seem simple enough but you can be sure there will be lots of devils found in the details. Usually to estimate R&D cost a team of analysts are paid a few hundred thousand dollars to perform a deep investigation typically taking several months but this analysis always takes longer and cost overruns are to be expected ...and so the cost analysis costs more than originally expected and when it's finally complete they come to some conclusion that a project will take x amount of time and money...but that project always takes longer than x as cost overruns are to be expected and so the cost is more than the cost analysis originally expected…so a performance review is initiated by a team of analysts who are paid a few hundred thousand dollars to perform a deep investigation typically taking about a year but this investigation always takes longer and cost overruns are to be expected. But I don't need to justify my MBA induced $50,000 student loan indebted, father disappointing existence- so we can take a shortcut and just estimate R&D cost as a historic average.

Saturn V: Cost - $83 billion, Years - 7
Space Shuttle: Cost - $46 billion, Years - 10
Falcon 9: Cost - $0.357 billion, Years - 4
SLS (Space Launch System): Cost - $19.1 billion, Years - 9
Starship (SpaceX): Cost - $5 billion, Years - 5
MMRTG: Cost - $0.083 billion, Years - 5
Perseverance Rover: Cost - $2.2 billion, Years - 7
Ingenuity helicopter (Mars Helicopter): Cost - $0.08 billion, Years - 5
Large Hadron Collider (LHC): Cost - $5.15 billion, Years - 15
Hubble Space Telescope: Cost - $5.51 billion, Years - 12
Manhattan Project: Cost - $30 billion, Years - 3
International Space Station (ISS): Cost - $175.86 billion, Years - 23
Curiosity Rover: Cost - $2.83 billion, Years - 8
James Webb Space Telescope: Cost - $10 billion, Years - 20
Tesla Model S: Cost - $1.7 billion, Years - 5
Google's Waymo: Cost - $1.18 billion, Years - 6
Human Genome Project: Cost - $5.28 billion, Years - 13
Airbus A380: Cost - $26.63 billion, Years - 15
IBM Watson: Cost - $2.11 billion, Years - 3
Blue Origin's New Shepard: Cost - $2.5 billion, Years - 14
Boeing X-32 (Joint Strike Fighter): Cost - $1.22 billion, Years - 4
Square Kilometer Array Radio Telescope: Cost - $2.62 billion, Years - 4
Apple iPhone (Original): Cost - $0.185 billion, Years - 2
Microsoft Xbox One: Cost - $3.29 billion, Years - 5
SpaceX Dragon Capsule: Cost - $2.81 billion, Years - 6
Toyota Prius: Cost - $1.68 billion, Years - 3
Oxford-AstraZeneca COVID-19 Vaccine: Cost - $2 billion, Years - 2
Boston Metals Electrolysis Reactor Cost - 0.352, Years - 12
Average Years: 7.66
Average cost per year: 0.513
Total: $3,929,580,000

Okay so the list actually isn't that long. I was going for at least 100 items spread across as many industries as possible but it turns out most companies aren't actually that happy to publish R&D numbers I guess because they want to keep secrets from their competitors. So most of these are government numbers or values revealed in court documents.

But I think it's good enough for our ballpark estimate. I could make it longer but- meh. With 28 items we get an average yearly cost of $1.658 billion with an average project duration of 8.1 years for an expected R&D cost of $13.265 billion.

But this is not a good model because things cost less to develop when they're no longer cutting-edge. There's a lot of basic research needed between 0 and 0.5, progress isn't linear, there's a steep initial learning curve. Saturn V development began just 5 years after we had put a single satellite into orbit. The Manhattan project began just barely 3 years after fission was even discovered. We are much further down the development track than that and aren't inventing anything from scratch, almost everything we are using is derived from existing technology. The most cutting edge ambitious thing is the excavator, but excavators and lunar rovers already exist; we are just combining the technology.

So if we remove the extreme outliers like the Manhattan project and the Saturn V and the International Space Station (which is the most expensive thing ever built in history by the way, mostly because the only thing more irresponsible with money than the government is 15 governments). I also didn't include anything Boeing, which you won't question if you know anything about Boeing. With these adjustments our new numbers are much more favorable with an average yearly cost of $513 million and an average project duration of 7.66 years for a total R&D cost of $3.929 billion. We might as well just round up to $4 billion over 8 years. So our costs were basically cut in half but our time remained the same. Obviously this will take longer and cost overruns are to be expected so you might as well hire your team of investigators now so they can finish investigating around the same time the project wraps up.

Now that we have our R&D cost, it's just a simple matter of adding up an estimate for all that equipment we just researched and developed. How much does a Lunarship cost? If you recall from the chapter ADDRESSING THE XLEPHANT we suggested a Lunarship build cost of $50 million, plus a lunar launch cost of $80 million. For ships-turned shelters that are not returning their usable lifespan is not 5 launches but rather just 1. It might be possible to do the retrofitting necessary after it has already flown 4 or more missions, thus distributing the build cost, however, let's assume this is not the case and it eats up the entire build cost on its first flight to the Moon where it will stay. Also since it doesn't need to return to Earth it would need less fuel and so its flight cost would be less than $80 million however we also wont factor this in just to remain on the high end of estimates. So we get a Lunarship-turned outpost cost of $130 million per unit. This is not its value, just its cost to build. We are maintaining its value by reusing it as a structure rather than a reusable launch vehicle. So we retrofitted, dropped, flipped and buried three ships, that's $390 million. We also transported two electrolysis reactors with other equipment on two different trips and we transported humans to the Moon once in our journey so far, the engineers who constructed the base but had to leave after two weeks due to radiation exposure. After this another crew will come and these will be the ones to furnish the place and since they can shelter in the radiation protected outpost they can stay longer and will oversee the installment of the reactors and machine shop etc. So let's add a second trip. Then let's assume all this takes place over the course of one year and that there is at least one crew change after six months, so adding another trip for three total. Then let's add a fourth trip for margins sake, maybe a crew resupply mission or something. So two electrolysis reactor payloads plus four human-oriented trips for $600 million. Our nuclear reactor cost is $10 million but was delivered aboard one of the ships turned shelters. To find out how much the two chromium anode reactors would cost I was going to compare them to the cost of similar reactors used in industrial aluminum production but I couldn't find this information without paying something like $2,000 for a report (ha!) but a quick look at industrial furnace costs shows they're typically between $10-30,000 dollars. I saw nothing over $500k so it's probably safe to assume these reactors will cost tens of thousands rather than millions which makes them a rounding error but to be safe let's just say they'll cost $1 million each. This should also be more than enough of a margin to cover the cost of all the auxiliary equipment that comes with them, the caster being the most expensive. So add another $2 million to our scorecard.

3x ships-turned shelters: $390 million
2 reactor payloads + 4 human trips = 6 flights: $600 million
1 nuclear reactor + 2 electrolysis reactors: $12 million
Total: $1.002 billion.

So $1 billion plus our R&D costs for a total of $5 billion over the course of 8-10 years. While $5 billion may seem expensive to a peasant like you, let me put it into perspective: we can gain the ability to develop the Moon for the same price as 45 F35 fighter jets or the Seattle Seahawks or 25 Marvel movies, or 2% of the money the government wasted in 2022 alone.

For comparison the US government wasted 247 billion in 2022, the Department of Defense currently fields 450 F35's, NASA spent $11.8 billion to develop the SLS, and the Pentagon spends $500 million a year to run Guantánamo Bay ($13 million per prisoner), and as of 2023 humanity has spent about $6.4 billion to produce 32 Marvel movies with 12 more in development; so $5 billion to begin developing the Moon is not much, all things considered.

PROFIT

Did you know the USS constitution was nicknamed "old ironsides" because the wood it was built of was so strong and dense cannonballs would literally bounce off it. This was because the ship was made from old growth trees. Trees hundreds of years old, great big white oaks and pine found only in the vast virgin woods of the New World.

Dense. Mature. Strong.

By contrast the old world, covered in vast farms, where man had felled the primeval forests a millenia ago, had now only specially designated royal forests left: all new growth. In a centuries long shipbuilding arms race with France and Spain, the British naturally began sourcing much of its lumber from the American colonies, providing them with a considerable advantage. By the mid 18th century the Royal Navy, so formidable, unmatched by any equal, was built from American wood. By the time of the American Revolution shipbuilding had become the lifeblood of New England. In the age of sail shipbuilding was one of the most capital intensive activities done on Earth. It doesn't make sense to build ships on the Moon today just as it didn't make sense to build ships in Massachusetts in 1550 despite Massachusetts having better wood... But by 1750? Once Massachusetts had a developed settlement and infrastructure building ships there became a logical conclusion and ship building was the northern colonies main economic driver because it had the natural advantage of hardy new world old growth wood. But what drove settlement and capital to New England before there was a thriving shipbuilding sector? New England is not known for its rich fertile soil and abundant crops. No, that was Virginia and Carolina. The incentives that began and grew the Plymouth Company colonies were intangible. Religious freedom, cultural freedom, the general pursuit of an unrepressed lifestyle drove people and capital across the atlantic. It was not some 'high value trade good' like with the less populated Virginia company settlements but rather something more transcendental. Less quantifiable, but possibly even more powerful.

But let's be real, even if the Moon was a tropical paradise and not a radiated hellscape I doubt religion and the pursuit of freedom would lead to its development today. If only there was *something* similar today, some sort of intangible thing people valued enough to pay for. But that's silly, I mean, afterall we all know modern economies are *just* manufacturing based right? Right?

Oh wait, what about *TOURISM?*

Imagine if you wanted to start an industrial technology company. It doesn't make sense to start an industrial tech company in a random field in Texas in the middle of nowhere. No engineers will want to move to a random field in the middle of nowhere because there's nothing to do and nobody around. Plus you'd have to construct a new building from scratch and since we're in the middle of nowhere labor and materials will have to travel further which will increase their cost. Then your facility needs water, power, and sewage which means you have to pay to install a powerline, pipelines, and a septic tank which will need to be cleaned annually, further increasing your costs.

But if that same random field was, over time, developed for some 'other' reason: perhaps it was found to contain a ton of oil or gold, maybe it was religiously important, or it simply had a nice river, then it might be transformed from a random field into a small city. Now it has all the places, parks, and amenities your engineers would like, plus easy to connect water and sewer, road infrastructure and power. Now it makes a lot more sense to start your tech company there.

For the Moon this 'other' thing that will develop it is tourism. Across the Earth tourists flock to exotic destinations, carrying with them resources and capital– the means to develop untamed land into habitable territory– and what reachable destination is more exotic and in need of capital than the Moon? As travel to the Moon becomes more accessible, people will pay to experience the awe-inspiring views, the thrill of low gravity, and the novelty of being on another world. Like the impact of tourism on the economies of developing countries, the boom in lunar tourism will spur the growth of the service sector. Tour guides will be among the first of these service providers, leading lunar excursions and providing visitors with a comprehensive understanding of the lunar environment. Tour guides, providers of goods, and operators of souvenir shops selling 3D-printed mementos will all find a market on the Moon. As the lunar base grows, so too will the demand for expansion.

New shelter units will be needed, offering greater comfort and amenities. Specialist roles will emerge, from repairmen of EVA suits and 3D printers to construction workers and rover mechanics. #MADEONTHEMOON. Mugs. Desk toys. Fidget spinners. T-shirts. Anything that can be 3D printed will, for a time, be coveted as novelty. Also, even Moon rocks and lunar regolith could be exported. Wouldn't you buy a lunar rock paperweight or a Moon mug? Zero to Low Gravity 3D printing is already well established and a printer by MadeInSpace is currently used to 3D print tools for astronauts on the ISS.

Other appeals include the ability to gaze upon the greatest night sky imaginable during the lunar night, unfiltered and unobstructed by an atmosphere or light pollution, an experience increasingly rare on Earth. The ability to play sports and a few other physical activities I'm too polite to mention– all enhanced by low gravity. Low-gravity paintball, racing, and UFC fights. Low-gravity dancing and yoga. Low-gravity trampolines and trapeze. Nearly anything you can think of that we normally do here on Earth but with 1/6th the gravity, which naturally makes it six times cooler. Once the seed of a space economy has been sufficiently nurtured by tourists, lunar development will organically grow as effects continuously compound, externalities increase, and local advantages layer.

So, if we want a self-sustaining presence on the Moon, one that justifies itself economically so humans are kept in the heavens via market demand and not just political forces or ideological aspirations then we need to host tourists and to do that we'll need to have a profitable lunar tourism base, a lunar resort (for lack of a better term) which… sounds kinda boring. I don't want to be a hotel owner as much as the next guy (no offense hotel owners reading this) but the goal here is to kickstart a lunar economy with the eventual goal of doing much more with it. If a New York City hotel owner can become president of the United States, just think what a lunar hotel owner could do. Actually, let's try to estimate profit potential.

Our flight cost to the Moon is $100 million, and our internal fairing volume is 1,000 cubic meters. SpaceX has said Starship would be able to carry 100 people to Mars, but that is a 7-9 month journey. The Moon is only about 3 days away. Considering A Boeing 747 has an internal volume of 876 cubic meters (less than Starship) and carries 416 passengers, 200 for a 3-day Lunarship journey to the Moon is reasonable. $100 million divided by 200 passengers is $500,000 each for transport break even costs. Unfortunately even after 42 years of trickle down Reaganomics not all 8 billion humans can afford to dish out a minimum of $500k for a trip to the Moon. So then how many humans CAN? I guess a good place to start is by asking how many millionaires there are. In 2022 the world's millionaire population amounted to 59.4 million adults. Credit Suisse forecasts that the number of millionaires will still grow to 86 million by 2027, a 45% increase from 2022. But not all people with at least 1 million dollars will buy a ticket. Some don't care, some aren't healthy enough to make the trip. Some people will buy multiple tickets while others will buy just one and many will buy none. But I think we have a metric that already exists involving people who have money, health, and an interest in things like space, science and tech. Something that would correlate well with potential ticket buyers: *General Aviation.* Like potential ticket buyers some people have one aircraft, other people have many aircraft and some aircraft are owned by many such as in flight clubs, but all aircraft have owners with money to spend on an interest that should overlap with lunar tourism. Sure this is comparing apples to oranges but at least they're both still fruit. Now, we cannot just count the number of aircraft in general aviation as most aircraft cost less than $200k and most general aviation aircraft are over 50 years old and the price of aircraft has risen significantly over the past 50 years even adjusting for inflation. So really we need to limit our data to aircraft costing over $500k that were sold recently. This limits us to turboprop airplanes, helicopters, and business jets sold within the last 5 years. This gives us a range from about $500k to several million. According to data released by the General Aviation Manufacturers Association, the total number of worldwide aircraft that fall into this category is: 10,763. That's nearly 11,000 people who spent at least $500k on an interest that would likely align with lunar tourism in the last 5 years. Since Credit Suisse estimates a 45% increase in millionaires over the next 5 years I think it's safe to just round up to 11,000.

So we have our target market and our transport break even costs but what about our build and labor break even costs? To estimate both these costs we need to know how long our guests are staying. Let's say we send up 200 guests at a time and they stay for an average of two weeks. Why two weeks? Well it doesn't have to be but I think it makes sense because of the Lunar cycle- moonwalks during the day, stargazing at night. And it's actually kinda long, don't you get tired after two weeks abroad? A higher turnover rate means less capacity so it also helps us remain on the high end of a conservative estimate. So 200 guests every two weeks would be 5,200 a year meaning we would be able to serve our entire target market in 2 years.

Some of the highest rated cruise lines have a crew to passenger ratio of 1:12 while the lowest rated ones have ratios around 1:25. Since we're serving the ultrarich let's use a ratio of 1:10. Also borrowing from cruise lines typically employment contracts last 6-months and due to the low gravity causing bone and muscle atrophy, 6-months on the Moon is a long time although many astronauts have stayed longer on the ISS which has basically no gravity. This means we'll have 20 employees living and working on the Moon for 6 months at a time, or 40 employees a year, 80 contracts over 2 years. If each contract pays $100,000 then we'll need to spend $8 million on wages, which would add $727 to each ticket. One lunarship can carry enough food to feed 626 people on a 3k calorie diet continuously for one year, and since we are hosting 200 people plus 20 staff (220) continuously a single launch-worth of food over the course of two years is more than enough, adding another $9,000 to each ticket. Note I am not suggesting sending up two years worth of food in a single launch; this is a launch-worth of food, most likely sent up with each trip. We also need to add the cost to launch the staff to the Moon to each ticket which would be 80 people or 40% of a launch ($40 million) so add another $3,636 to each ticket.

How large do we need to build to host 220 people at a time? Well Cruise ships have about 122 cubic meters per passenger and the international space station has about 130 per astronaut so let's go with 130 cubic meters per head which would result in us needing 28,600 cubic meters of habitable space. Let's round up to 30,000. To build this large we must deliver another nuclear reactor since it only provides enough power for 20,000 cubic meters. Two reactors would provide enough power for 40,000 cubic meters and with our first outpost (+1,000) and the machine shop (+1,000) and the first +1,600 cubic meter building plus the 30,000 cubic meters we intend to build we'd have 33,600 cubic meters needing power. So let's deliver the second reactor with a second excavating rover aboard a ship that will land sideways, open like a cargo-plane and unload the rover, then close and be buried to serve as the reactor housing same as we did the first one. This whole process adds another $140 million ($10 for the reactor + $130 for ship). Our steel production rate is unknown but 30,000 cubic meters isn't a small project and we still have a bunch of empty space in the rear of that second ship so we might as well send up another two electrolysis reactors for an additional $200 million. Now we've essentially doubled our build rate capacity (unknown) for an additional $340 million, adding $30,000 to each ticket. As for construction labor cost we don't know if most of this can be done remotely using Boston Dynamics Robots or if humans need to be present or just partially present. This cost could only be estimated by simulating the entire build process and finding the optimal strategy first, but let's just assume 2 years of build time involving 4 flight missions and a team of 20 robots/humans being paid $150,000 a year (Boston Dynamics Atlas cost about $150,000). This would total $106 million or $3,533 per cubic meter, adding a final $9,636 to each ticket.

Our final ticket break even cost is: $552,999.
Obviously we have to round that up to $553,000.

Importantly we did not push our $4 billion R&D costs onto our 11,000 consumers for several reasons. First it is safe to assume much of this cost would be subsidized by grants from Uncle Sam. Second, even if that were not the case the R&D costs directly translated into the ability to *build on the Moon*. This is very powerful and can generate revenue in many different ways outside of tourism. I'm sure NASA and ESA and JAXA and CERN and the Pentagon and many other private corporations will be lining up to pay *you* to build them a cool lunar crib. Our tourism analysis is just to get a very narrowly defined conservative baseline estimate of earning potential to prove the viability of being a lunar hotel owner. We only added up the ticket cost for things directly related to tourism. So with a $553,000 break even ticket cost the only question now is how much should we charge for profit? The higher the ticket price the more people will be priced out lowering our earning potential, but we also have to consider demand as the first trips will be worth more than later trips and the best way to do that is to auction off the tickets- letting the market decide.

However, just to get an idea: if the average ticket sold for $600k we'd profit about *$517 million* in 2 years.

WALMART

You can do A LOT with steel panels and beams and five axis cnc machines, yet no matter how advanced your domestic lunar production capacity becomes, no matter how much you make off world, this roaring economic engine will always need inputs from Earth or elsewhere for anything that cannot be easily sourced on the Moon. No organic compounds, none of our favorite breathable inert gasses, no edible carbohydrates and hydrocarbons like maltose and microplastics; and- realistically- no meaningful amounts of water for anything larger than a few dozen people. This shouldn't be that surprising, every modern economy on Earth- even the most insulated still need imports. But how much our lunar economy can import will depend on two things: How often we can land a load on the lunar surface and how large that load is.

A Lunarship-sized super-heavy lift launch vehicle has a generously high-end estimated payload capacity of 200 tons to the lunar surface. If we land 1 Lunarship-sized rocket on the Moon every single day, 365 days a year, then we could deliver 73,000 tons of stuff a year. That may sound like a lot but it's not. It's not enough for even a small lunar economy, much less a growing and thriving one. To put it into perspective that's less than a single medium-sized panamax bulk carrier from 1980 can carry, which is about 75,000 metric tons a year... wait, no- not in a year.

In a single trip.
At any given time.

But okay I realize comparing cargo ships to rockets is like comparing apples to atom bombs, however we're not concerned about energy but rather logistics here and 73,000 tons a year is very little, so let me try another comparison that's probably a bit more familiar. A small Walmart receives about 5 trucks a day to remain stocked which would be about 1,825 trucks a year. This was surprisingly hard to find data on but according to several truck drivers it seems that the typical semi-truck load weight is roughly 22 metric tons in the US, although it can be much higher, but let's just say 22 tons per truck as a conservative estimate meaning a single Walmart receives about **40,150 tons** of goods a year.

I'm basing this on info garnered from Quora and Reddit posts... but after all 7,000 grains of barley weighs 45% as much as a chunk of platinum-iridium alloy from 1889 so I am officially proposing the addition of a new unit of measurement to the International Bureau of Weights and Measures: the "Walmart" which is equal to 40,150 metric tons, 40,150,000 kilograms or about 619 billion grains of barley. If we land 1 Lunarship-sized rocket on the Moon *every single day*, 365 days a year, then we could deliver 1.8 Walmarts worth of stuff a year. But how realistic is that launch frequency? I mean can we actually land 1 Lunarship-sized rocket on the Moon every single day of the year?

Well, no.

2023 saw 223 orbital launches, setting a world record for the third year in a row. That averages out to roughly a single launch every one and half days across all of Earth. Think of all the things that go into a single launch from Earth, the logistics, communication and supply lines, the timing and tracking and coordinating. All of this, done once every 24 hours on Earth by an essential workforce with rare skills and experience who need vacation and sick days. And now realize that launching a rocket from Earth is not the same as landing a rocket on the lunar surface, and this landing aspect is likely the tighter end of this bottleneck as you have to take all those things involved in an Earth launch, and do them again for landing on the lunar surface, plus the extra steps of unloading, refueling, and then launching again from the moon back into orbit, all in under 24 hours. This means a round the clock workforce with constant supply inputs, not to mention that any time delays and scrubbed launches will cause a traffic jam in the logistic network both on Earth and on the Moon. And there's solar storms and meteoroid streams which is sorta like a space weather problem of its own. So landing a fat load of any size on the lunar surface once every 24 hours is likely impossible in both the near and mid term.

But let's say we can, optimistically, get to the point where we're able to launch a superheavy rocket from Earth every single day of the year, 365 launches a year. These would not be direct trips from the ground to the lunar surface, because in reality Starship-like vehicles need to undergo orbital refueling once reaching low earth orbit to give them enough juice to make it the rest of the way to the Moon. It's currently estimated that for each trip to the Moon, anywhere between 5-12 fuel launches will have to be performed.

At a 10:1 ratio, for every ship sent to the Moon, 10 fuel launches would need to be performed, 11 launches total. At a launch rate of one a day, out of 365 launches a year only 33 would be Moon-bound yielding just 6,600 tons, or 16.4% of a Walmart. With 8 fuel launches per trip we'd net 20.2% of a Walmart and with just 5 launches we'd net 30.3%. But let's be super optimistic (and unrealistic) and say the ratio of payload to fuel launches is just 2:1, so for each moon bound ship only 2 other fuel-ships will need to be launched, 3 launches total. At this unrealistic ratio, if we launch the largest and most powerful rocket ever built, every single day of the year only 60.6% of a small Walmart would make it to the Moon.

And then we still have to get people there! So far we've just assumed people are on the Moon but they have to get there using the same vehicles as the goods they're consuming, sharing from the same pool of 365 launches a year! Using our unrealistic ratio of 2:1, out of 365 launches in a year 244 would be devoted to refueling missions. That leaves us with 121 launches for goods and the humans they're intended to supply. So if we can only get 60% of a Walmart to the Moon, we need to know how many people that can support, and how many launches it will take to get that many people to the Moon and those launches will reduce the percentage of a Walmart which will reduce the number of people and launches, so we need to find the optimal ratio of people: to cargo launches. The first step is to find out how many humans would a single 'Walmart' support? And we're not talking about survival necessities- food and water. We're talking about a lunar economy with a large tourism base- so *amenities*. Sure in the case of tourism they're there for the experience itself and people often live in smaller quarters on cruise ships than they would otherwise, but cruise ships still have alcohol and games and stuff. So how many people can a Walmart worth of amenities support? Well "people-per-Walmart" is kinda a bad question because man cannot live off Walmart alone, there are lots of other suppliers in the market so how could you ever find a reliable metric to base this on, I mean it's not like there are any places that just have a single Walmart providing for the entire community like some corporate dystopia deserted island-Walmart scenario...

Oh wait, that's the US midwest.

All across the midwest and Appalachia there are small towns with populations between 2-6 thousand people whose entire economy and existence is supplied by a single Walmart. Most of the town works at the Walmart and shops at the Walmart. Like mining company towns of old, they are Walmart company towns. Which kinda confused me. How are these sustainable if they don't produce anything, they just consume Walmart goods... in exchange for working at Walmart? Sure coal-company towns monopolized the entire market, but they existed to export coal. They produced *something*. I grew up in a small town in Texas, but it was near a large city, and most people commuted to the city for work. But these towns I'm talking about are in the absolute middle of nowhere, way too far to commute. So then how do these towns, which don't commute to a nearby big city, and don't really seem to produce anything, no factories or workshops or mines, not even farms in most cases, remain on consumption alone?

I'm making broad strokes here as every one of these forgotten places has a unique background and character but in general it seems that most of these places were established and grew to their peaks from two things: Energy or Transportation. Either coal or oil being discovered there at some point a hundred years ago, the mines and wells now long since exhausted, or a railroad or cattle trail stop.

And that's pretty much it.

But that's the story for tons of abandoned ghost towns.
Why are these still here? How are these places sustainable?

Well the answer is: they aren't. All of them have seen population declines every census for decades except in 2020 when many moved from the city to the country during the Covid-19 pandemic, but this was largely *temporary* and population declines returned the next year. It seems most kids who grow up in these areas move out as quickly as possible. So they are on their way to ghost town. They are dying, just slowly. Does having a Walmart slow their death spiral? Well there is only a Walmart because Walmart felt there was a large enough serviceable market to warrant the upfront investment so there has to be *something* other than Walmart bringing money into the town. In fact if the Walmart is making a profit it is technically siphoning money from the town. So then where does the money come from?

Those who remain and aren't employed by a local Walmart are typically employed in either trucking or the oil industry, the nature of which sees them leaving home for several weeks at a time and then returning for a few weeks of off-time. So they do commute, but instead of heading to a nearby city, they traverse state lines and national borders. Interestingly, this pattern of employment links them to the very industries that originally led to the establishment of their towns: Energy and Transportation. These industries have deep historical roots in the community, shaping its identity and economic foundation over generations. Deep roots to the past. Roots, which give these places staying power long past their expiration. Debt, welfare, addiction- they all do their part in keeping people in places like these, but the real staying power, that which has kept these towns on a map with enough population to warrant a Walmart are their roots to the past. Like old oaks communities with deep roots may stand for generations, nourished by the past. But while roots can be nourishing, they can also be entangling, entrapping, suffocating. While many may leave, emigrating to greener pastures, places of growth, these forsaken places endure because every year some of the youth remain, some who might otherwise have found opportunities and fulfillment elsewhere. Trapped by their families, their memories, their homes. The weight of history both anchors and restrains. Every generation that remains sustains the community, preserving its traditions and stories.

But the truth is, eventually, the day will come when even the strongest oaks must bow. These towns are nothing more than the high water marks of a long forgotten wave, the remnants of historical echoes in a void, silently living out their twilight years, quickly approaching oblivion as one by one homes are left to rot and the concrete cracks grow. No matter how many once called them home, eventually these forsaken places will fade.

Even the oldest trees, with the deepest roots are toppled eventually; usually cut down to make way for a new Walmart.

But we still don't know how many people a small Walmart-worth of amenities can serve per year!

2145 Eastern Ave, Gallipolis, OH 45631	Pop. 3,297
100 McGinnis Dr, Wayne, WV 25570	Pop. 1,405
4490 Gallia St, New Boston, OH 45662	Pop. 2,245
240 Wal-Mart Way, Maysville, KY 41056	Pop. 8,742
11217 State Rte 41, West Union, OH 45693	Pop. 3,002
1750 S Perryville Blvd, Perryville, MO 63775	Pop. 8,488
6495 Country Club Rd, Murphysboro, IL 62966	Pop. 7,033
1870 W Main St, Salem, IL 62881	Pop. 7,113
2700 W Broadway St, Princeton, IN 47670	Pop. 8,367
2251 IN-54, Linton, IN 47441	Pop. 5,162
1304 E Main St, Robinson, IL 62454	Pop. 7,094
1701 N Main St, Beaver Dam, KY 42320	Pop. 3,490
1725 W Everely Branch, Central City, KY 42330	Pop. 5,810
1600 W Main St, Walnut Ridge, AR 72476	Pop. 5,492
1007 N Douglass St, Malden, MO 63863	Pop. 3,580
60 S Stewart Rd, Corbin, KY 40701	Pop. 7,856
175 Walmart Plaza Dr, Monticello, KY 42633	Pop. 5,755
1650 Edmonton Rd, Tompkinsville, KY 42167	Pop. 2,297
2988 Burkesville Rd, Columbia, KY 42728	Pop. 4,846
333 E Walnut St, Thayer, MO 65791	Pop. 1,900
1433 S Sam Houston Blvd, Houston, MO 65483	Pop. 2,163
1888 Hwy 28, Owensville, MO 65066	Pop. 2,806
1802 S Business 54, Eldon, MO 65026	Pop. 4,545
1250 W Dallas St, Buffalo, MO 65622	Pop. 3,422
3020 S Elliott Ave, Aurora, MO 65605	Pop. 7,412
500 W Mt Vernon Blvd, Mt Vernon, MO 65712	Pop. 4,550
2115 S Main St, Grove, OK 74344	Pop. 7,215
2121 TX-16, Graham, TX 76450	Pop. 8,750
2614 N Swenson St, Stamford, TX 79553	Pop. 2,921
2801 Avenue F, Childress, TX 79201	Pop. 5,797
12910 OH-664, Logan, OH 43138	Pop. 7,258
200 Wal St, Summersville, WV 26651	Pop. 3,369
1983 S Mississippi Ave, Atoka, OK 74525	Pop. 3,200
2008 W Grant Ave, Pauls Valley, OK 73075	Pop. 6,026
1907 E Washington St, Idabel, OK 74745	Pop. 7,056
772 Airport Rd, Cleveland, OK 74020	Pop. 3,217

This is a random smattering of 36 Walmarts in towns across the midwest and Appalachia that have less than 10k population and aren't located close to a major city. I literally just typed into google maps "Walmart" then zoomed into the middle of nowhere, then clicked "search this area" and did that 36 times. There are a LOT more (obviously) so this is neither rigorous nor exhaustive but it gives us a general idea of how many people we can have per Walmart.

The lowest population was 1,405
The highest was 8,750
And the average was: 5,075

Let's just round that down to 5,000 people per Walmart. We also have to consider this a high-end estimate because Walmarts typically get their nitrogen-oxygen atmospheres for free from the planet they love to pave over, and thus they don't have to maintain an enclosed and pressurized environment with an extensive and material-heavy life support system that, even with very high efficiencies will always need some level of input, all of which eats into our annual mass allowance. In other words there's a lot more overhead, as some of the Walmart-worth of mass will need to go into maintaining the figurative Walmart itself... but anyways we'll ignore that and just say 1 Walmart equals 5,000 people. So of our 121 lunar-landing launches how many should be devoted to people and how many to cargo?
For instance if all 121 were devoted to cargo and the humans were left to hitchhike to the Moon then we'd achieve that 60% of a small Walmart discussed earlier, which would be enough amenities for 3,000 people. But since humans cannot hitchhike to the Moon, how many of those 121 launches would it take to deliver 3,000 people? How many people per launch?

A Boeing 747 has an internal volume of 876 cubic meters and carries 416 passengers. A 200-ton capable Lunarship-sized rocket would have a 1,000 cubic meter fairing, 14% more than the 747, but would also have to carry more supplies per passenger on a much longer 3-day journey to the Moon than the 747, so we'll call it even and say a single launch to the Moon can carry 400 people.

So at this point we need to figure out the optimal ratio use of our launch allowance, which is recursive. At 400 people per launch it would take 7.5 launches to deliver 3,000 people to the Moon, Which would subtract from our 121 launches, leaving us with only 113 launches for supplies which would reduce the number of people we can support from 3,000 to 2,814 which would require 7.03 launches which would leave 114 for supplies which is enough for 2,839 which would take only 7.09 launches.

So we could host about 2,800 people continuously. That is the maximum growth bottleneck of lunar development, launching a fully reusable 200-ton capable super heavy lift launch vehicle Every. Single. Day. That- unrealistically- only needs *two* orbital refueling launches.

When we imagine humanity's future in space do we imagine 2,800 people? Or ten thousand? Ten million?

So that's the problem. The lunar economy's growth will be bottlenecked- not by capital investment or even launch costs- but by import logistics: launch frequency. More than anything else. Sure you can establish new launch sites and expand old ones to try and meet demand and loosen FAA standards but every scrubbed launch due to weather or technicalities will cost time and money, building out lots of infrastructure, further increasing scale and supply chain logistics for everything from fuel to ground operations. We can also increase the size and payload capacity of the rocket itself, building an even larger rocket than those currently in development. Besides the additional upfront R&D costs this will also require an increased buildout of launch infrastructure, larger facilities for larger rockets, more scale and supplies running through larger and longer supply chains and a larger workforce. And actually this sounds great, a massive thriving space industry employing hundreds of thousands of people is a future I want, but each of those steps would only marginally increase the percentage of a Walmart delivered per year and we already took a highly optimistic measure of 2 fuel launches per trip, almost all improvements we could do on the ground in the next few decades would just get us to that level.

Now I know it seems crazy that saying 2,800 people on the Moon isn't enough, at a time when we haven't had a single person on the Moon in over half a century, but unless we want humanities lunar presence to end up like those Walmart towns in rural america, a flash in the pan, then we'll need to do something drastic to increase our supply lines- and with it our off-world economic staying power.

INFRASTRUCTURE

Before we had rails and roads we had rivers. Unfortunately the Moon has none of these things so we have to do everything ourselves (*as usual, grrr*). We need to move a lot more mass between the Earth and the Moon but unfortunately the go-to answer of simply increasing the size of the rocket would introduce a ton of complications for only marginal gain. It's possible to do this for sure, but it doesn't really solve the core issue. The real issue impacting lunar-landing frequency is not the size of the rocket but rather the unavoidable need for orbital refueling of superheavy-lift launch vehicles if they're going to make it to the Moon.

But what if they don't?

What if instead of going all the way to the Moon they could rendezvous with another spacecraft in Low Earth Orbit which could take on their payload and fly it all the way back to the Moon? Well then the Starship-like rocket wouldn't need to be refueled, it could just serve as a ground to Low Earth Orbit shuttle, and the secondary spacecraft would never need to enter into Earth's atmosphere which means it won't need to be aerodynamic and if it's just taking on cargo it can be fully automated which means no humans which means no pressurized interior hull- it can be made very simply- just a platform which means we could make the entire thing out of steel made from lunar regolith using chromium-anode electrolysis reactors which means we can scale this thing to be extremely large and it could take on several 200-ton payloads in a single trip like how this section took on several paragraphs worth of ideas in a single sentence.

What if we make something like the shipping container of space? On Earth shipping-containers have revolutionized supply chains as they allow loads to be easily transferred from factory to truck to ship in a standardized fashion, so what if we could make something like that but large enough to take on the 1000 cubic meter volume of a superheavy payload. And what if, to facilitate this payload transfer between craft, we could build some sort of mechanized scaffolding- essentially an orbital port- just a simple matrix which could move payloads from the Earth-ship to the Lunar-ship? Maybe we can divide this port into three sections: One side for empty containers, the other for full containers and a loading area. Perhaps we can design it so that once a ship arrives at the orbital port it won't need to dock, it can simply maneuver into approximate range, open its fairing doors, and the payload could then be captured by the port itself; after which it would be transferred into a waiting container. That container would then be moved into the full area where an incoming freighter would pick it up after dropping off its load of empty containers on the empty side. Then that freighter would return to the Moon fully loaded with supplies for a growing and thriving lunar economy.

Then I suppose we'd solve our supply problem.

What we essentially just did was reinvent bulk-carrier ships and ports but for space. However one little detail we gleaned over was the actual return and landing of the freighter to the Moon. You'd think it would be as easy as just landing on the Moon, unloading, then launching back and repeating the process- but here our real troubles begin, for in those details, there be devils. You don't want to land the freighter too close to the lunar base for safety reasons, but you do want to unload your cargo close to the base for efficiency reasons. You also need to figure out a way to refuel the vehicle which means you need a fuel depot but you don't want that to be anywhere near your landing spot for explosive reasons and also you'll want to load your freighter with empty containers to return them to the port for that closed loop container trade.

So what, do you land away from the base and use a bunch of gantry cranes for everything? Do you load its cargo onto a rail cart? Do you add wheels to the freighter adding to its weight? Do you grab it with chopsticks? Well what if you land the freighter on a landing pad far removed from the base, and that landing pad itself can move the freighter towards the loading unloading and refueling area near the base entrance- sorta like the NASA crawler used to move rockets into place. And what if we gave the freighter a little boost in its liftoff by accelerating the landing pad to a few dozen Kilometers per hour by simply making the rails longer.

But why stop at a few Kilometers per hour? Why not go up to a few Kilometers per second and use the landing pad itself to also launch space freighters from the lunar surface back into orbit, so the freighters themselves wouldn't need to carry the dead weight of lunar-launch hardware and fuel? We could accelerate to orbital escape velocity using electricity so the landing pad is like a reusable catapult if we space two rails apart for each leg and run an electric current through the rails and then the landing pad legs ride upon a conductive armature which connects the circuit: **a rail gun.**

Lunar escape velocity is 2.38 Kilometers per second and $d = v/2a$ so if we maintain a gradual acceleration of just 1 G (that of regular Earth gravity) then our track length would need to be about 300 Kilometers and it would take 4 minutes to reach escape velocity. At an acceleration of 2 Gs our rail length is halved to just 150 Kilometers and 2 minutes and at 3 Gs it would be just 100 Kilometers or 62 miles for only 81 seconds. While 3 Gs is perfectly fine for cargo we may want to use this same system to launch people and 3 Gs can be a lot on the human body, but it can be withstood for short amounts of time. For reference astronauts on rockets experience it for up to a minute while fighter pilots experience 6 and even up to 9 Gs for a few seconds, however considering tourism would likely be a major part of our lunar economy it makes sense not to limit our visitors to only highly trained fighter pilots. Legally roller coasters can pull 3 Gs for no more than 12 seconds so a passenger version would likely need to be about 300 Kilometers although we could shorten this by having a gradual increase in acceleration and accelerating more towards the end- pulling 3 or even 4 Gs in the last few seconds. Initially it makes sense to just use this for cargo per the cargo-freighters and secured cargo can easily accelerate at 4-5 Gs or more without much of an issue. 5 Gs would give us a track length of just 25 Kilometers or 15 miles and would achieve escape velocity in just 48 seconds. But even 25 Kilometers is a significant distance. It would be by far the largest thing ever built offworld, however it can be made of very rudimentary components, just simple aluminum rails created from in-situ resources using a robotic rover construction fleet either automated or teleoperated from Earth and it would be worth the investment because, by saving the freighter from also having to accelerate itself up to escape velocity we save a lot more capacity for cargo which means more resources for more people for more habitats for more freighters for orbital port expansion for more cargo for more resources for more run-on sentences.

What we have just arrived at, through a bottom up-step-by-step problem solving approach is the creation of a reusable, electric-powered system used to drive mass into orbit.

A market-driven mass driver.
But it has issues.

The concept of the rail gun is over 100 years old, first conceived of in France during world war I and everyone from the Nazis, to the Chinese, the Russians, Indians and both the U.S. Army and Navy have all poured significant resources into the effort at one time or another with the Japanese most recently picking up the mantle. Despite all this no railgun has ever been fired in anger, no designs mass produced. It has never been used in war for the same reason it's considered a "bad" lunar mass driver design. *Ablation.* Essentially rail wear, gouging and grooving due to the high stresses imparted into the rails from heat and arching and forces involved in accelerating a projectile up to 2-3 Kilometers per second within just a few meter barrel. Ablation is the major bottleneck limiting the systems reusability. The US Navy was able to get off 400 rounds before a barrel needed to be replaced and was approaching 1000 before the program was shut down in 2021 but the major costs involved make it impractical as a weapon. But we aren't trying to build a weapon. We are trying to build infrastructure. All previous research has been done in the context of trying to make navy guns with super *short* track lengths and super *high* acceleration rates- like tens of thousands of G's. Very demanding. Lots of ablation. We have a lot more room to work with as we can make our track long and our acceleration rate low.

Despite this though we will still have ablation, and this is why railgun designs have largely been considered "bad" candidates for lunar mass drivers in favor of coil guns which suspend their payload between copper coils allowing for no surface contact and no structure wear. But the issue with this approach is the payload size would be constrained to the diameter of the coil-gun mass driver which means it won't work well with our freighter design, a design which was logical in being a basic easy to construct platform meant to simply move cargo back and forth. The only reason we reached mass driver was as a means to amplify the effectiveness of these freighters which themselves were the solution to our original market driven supply problem.

On the Moon and beyond Earth in general construction costs are the largest costs, which is why we need to put premiums on simplicity like with the freighter design, and the coil gun mass driver would have much higher construction costs derived from a more complex architecture demanding more resource intensity per cubic meter compared to that of the rail gun which is basically just simple suspended rails and capacitors. Now sure with the coil gun you have rapid reusability, no repair needed, and so while your payloads have to be smaller you can launch a ton of them in quick succession. But what I think most people miss is we aren't launching bullets, we are talking about launching spacecraft. You cannot just shoot chunks of the Moon out of this thing to a predetermined destination. The loads still need to be powered. There's all kinds of perturbing forces acting on any object at any given time. Consider the sun and moon and earth all interact which is why some Lagrange points are unstable. Even the varying topology of the nearest planet, different density at say mountain ranges vs maria has an effect when an object is within the hill radius. So any payload will need to perform descent burns to get into a specific altitude/orbit. We are talking about launching spacecraft which have to be manufactured and fuelled, and so the mass driver really needs to be thought of as a first stage for a rocket more than anything else and the same principles for why big rockets are better still applies. Scale.

So with this context in mind the question of railgun vs coilgun becomes: do you want to launch 10,000 tons aboard 2 large spacecraft or 10,000 tons aboard 1,000 little spacecraft? In either case it makes sense to build those spacecraft on the Moon and it's much easier to build two large spacecraft than 1,000 little ones.

As if that wasn't damning enough the 1,000 spacecraft, basically little rocket pods, have to be recaptured and transported back in a closed loop cycle. Are you just throwing them away? This means they also need to be decelerated back to the lunar surface somehow... man if only there was a big platform to capture lots of mass in orbit and transport it back to the Moon and then decelerate it.

Oh wait, that'd just be the freighter! But instead of transporting cargo in it's containers it would have to transport these little spacecraft, which are too small to be used to import a starship sized load unless you want to redesign the entire orbital port to be a much more complex sorting facility to piece apart every incoming payload into a hundred different bundles to fit into these pods which is stupid. A potential solution could be to make the pods much, much bigger so a single one can launch several starship payloads worth, and maybe you could land it on a landing pad that moves, and maybe you can move that landing pad really fast on some electrified rails to help with the first stage... and so you can see how a railgun design is optimal.

But we still haven't solved that ablation issue.

I am going to propose a radical solution to that.
What if, and hear me out, what if...

We simply *embrace the ablation.*

What if we simply embrace the fact our rails will be damaged and just simply repair between launches. So far we've talked about building nearly everything out of steel, the shelters and ships and rovers. Why bother with aluminum: the balsa-wood of metals when you have the option of using steel? But for the mass driver rails aluminum actually makes the most sense because aluminum has higher electrical conductivity than iron and even though we are embracing the ablation, like alcoholism we don't want to embrace it... too much.

Over time we can reduce this wear by upgrading rails with cooling systems and composite platings etc. By cooling down the rails to 50K or -223 degrees Celsius, -370 degrees Fahrenheit, we could reduce resistance by an order of magnitude, and this sounds like a lot but remember the lunar night already drops down to -133°C, -208°F naturally. So with all this in mind one could easily conceive of a lunar construction rover built around an electrolysis reactor that can print these rails. Regolith goes into one side, rails come out the other. But this is admittedly an oversimplification, really you'd need a fleet because the rails themselves should be on a surface that isn't just plopped down onto the moon because the electrostatic dust will interfere with launches. Fortunately you can lightly melt regolith into what is essentially a rough asphalt paste using orbital mirrors to concentrate sunlight or by creating the evil moon-version of a Zamboni- a nuclear powered melting rover. It doesn't matter whatever is easier.

Also the lunar terrain isn't nice and even, it's rugged and boulder strewn so you also need, basically, a lunar bulldozer to clear the path. Ya know, basic construction stuff.

Let's jump back to the mass that's actually being driven by this mass driver: the freighter. The freighter itself is just a simple platform, a powered carriage built to carry containers, we only called it a freighter because it was carrying freight. But what if you carry a tank full of liquid oxygen? Now is it a tanker? I dunno, I hardly know her, but I do know it wouldn't be hard to make pressurized tanks; you're already making pressurized habitats for people; this is the same thing, we just call it a tank when it holds something other than humans. If it holds fuel it's a fuel tank, if it holds fish it's a fish tank, and if it holds humans it's a lunar habitat. Unless it's getting shot at, then it is a tank. So what do you call it when you strap a human-tank onto this freightless freighter and then use it to take people from leo to luna the same way we did cargo, cutting out the 8 or so refuel launches per trip? Do we call this a ferry?

How much would this *ferry* reduce launch costs and how much more revenue could we generate from it? Keeping our Starship flight numbers, we'd have a LEO flight cost to rendezvous with this ship of about $20 million, then an unknown cost from LEO to the Moon using the freighter. In other words we know how much it would cost to get from the ground to low earth orbit but we don't know how much it would cost to fly from low earth orbit to the Moon on a ferry/freighter. How much did it cost to research and develop and build? How large is it and how many trips can it make during its useful lifetime? How much did the orbital port cost? The mass driver?

Fortunately we do know each ferry/freighter flight would be at least less than $10 million which is the projected marginal flight cost from the ground to Low Earth orbit of a Starship-like rocket which travels through an atmosphere, something the freighter is not doing- that's the whole point. In all likelihood it would probably be less than $1 million per trip because it should have a very long lifetime since- again- it's not getting beat up flying through an atmosphere, it just peacefully floats between the Moon and Low Earth orbit. But let's just go with ~$1 million per trip as a conservative estimate to account for amortized R&D and build costs. This would result in a transport cost to the Moon of ~$21 million.

Using our aircraft data once again this would open up our target market to include piston airplanes and would increase from 11,000 to 18,000 (17,959). But the only reason we used aircraft data in the first place was because we wanted a correlating interest with spaceflight in a market where people could afford the outrageous price of a ticket so we counted recent sales of $500,000 aircraft. This is why we didn't just count millionaires because even they couldn't afford a ticket, only the ultra-rich could. Also we limited our number of people per flight to just 200 since it is a 3 day flight but if we are able to accommodate many more people on the freighter then we can reduce ticket costs even more. We could build this thing to carry a thousand people but if it carries just 400, which is still less than a Boeing 747, then the cost per ticket would be reduced to just over $50,000 which means that- actually, we can just count millionaires. Polls have shown that 35% of U.S. adults say they would be interested in orbiting Earth in a spacecraft, so if 35% of people with at least $1 million opted to purchase a ticket, the serviceable market would be over 30 million people. With 30 million people purchasing a $50,000 ticket, if we just add a 1% profit margin which would only add $500 to each ticket then we'd have a profit- not revenue- but profit potential of over $15 billion.

Why stop there?

Once you have a ferry/freighter and mass driver, using them exclusively to increase lunar imports and profit is... like inventing a teleporter and only using it to advertise chocolate. If the Moon is humanity's gateway to the stars, a lunar mass driver is the key to unlocking that gate. So what if we assemble a bunch of these human-tank ferry hulls (habitable spaces) together into a big ring in lunar orbit? Now are they prefabricated station modules assembled into a big-ass rotating space station? and what if we use a few freighters to move a few of these big boys from lunar to low earth orbit, carving out significant amounts of habitable high-value real-estate, is the freighter now a cis-lunar tug? The station now an orbital reef? What if you replace the standard engines on these tugs with very powerful open cycle gas core nuclear engines and then use them to accelerate several stations fully loaded with people and supplies past Earth-escape velocity, into a Mars transfer orbit?

Well then, I suppose we'd call that a colonial fleet made from the Moon full of people made from the Earth.

So if you can do all this, if you can create modular, standardized, scalable, interchangeable designs, then it would be the end of the beginning of humanities expansion into the rest of the solar system.

THE ENTROPIC ABYSS

At the end of World War Two, the world was ordered according to two different visions: that of the US and the USSR. The Age of Imperialism was done. The Age of McDonalds and Tenement Housing had dawned.

In the West golden arches stretched across the Atlantic, heralding from the new world to the old:

"Rejoice for free market capitalism has arrived! And to celebrate this new age the US is offering a limited discount on all reconstruction of war-ravaged cities! All you have to do is sign up with the nearest US embassy and use promo code: 'Marshall' to claim this amazing limited-time offer! Terms and conditions apply"

In the East, an iron curtain rose under a red star, within it a new world would be made out of the old:

"Comrades of the world, unite! As a testament to our shared destiny, the USSR introduces a dedicated program for rebuilding lands torn by imperialist wars. Pledge your solidarity with the socialist cause at your nearest Soviet embassy and invoke directive number: 'Comintern' to participate in this grand endeavor! Collective duties and commitments apply."

This 'system' was not invented in the US, or by Americans, but it had been perfected in the US as a way to forge a new continent into habitable territory. The system led to the development of the North American continent, and then it re-developed the European continent after it had been reduced to ash. To win the cold war the system was expanded globally via what grew from a Bretton-Woods seed. You know it as globalization. This system has the amazing ability to create growth, but it also depends on frontiers to grow into. Its final frontier opened when the Berlin Wall fell. That was 12,405 days ago (at the time of writing) and since then globalization has devolved into a bloated frontier-less mess in which the stock market depends on the ability to control the entire planet to leverage itself back from the brink of collapse.

We now live in a global organism made of humans, their creative brains and nimble hands, the organs; material wealth, the blood. Like the most basic organisms, it is simply a basic feedback loop that grows through consumption. Remove the frontiers and the organism starves. It's no wonder then why the middle class is shrinking as the global organism within which they live undergoes autophagy.

Growth depends on a growing demographic, the basic unit of productivity, and the basic unit of wealth generation. Today in our developed societies we can have as many children as we want and they won't die. Yet we don't have many children today because we have to pay a price in terms of our social class. Everybody in the upper-lower to lower-upper classes today has a limited number of children because having more of them would result in a big economic hit to the family since children, especially in the US, are very expensive to raise. One is only taken out of this if they are on the extremes and are super wealthy or super poor. If they are impoverished it doesn't matter how many kids they have because their status is not going to change. In fact, for most of human history, when we were mostly agrarian communities, children were useful labor. Having more resulted in more productivity per household, so children were actually wealth-generators, the inverse of today. If one is super wealthy they can afford to have as many children as they want just like flat screens, cars, and boats.

Today the majority of families reproduce under the population replacement rate which is 2.1 as it takes two people to make one baby but one baby only replaces a single person. If we want our population to grow we have to have more than two children per couple. All developed nations except the US, France, New Zealand, and Argentina are below the replacement rate of 2.1, and both the US and France are only at the replacement rate after adjusting for net immigration. South Korea has the lowest, with a rate of 0.9 children per woman. As for developing nations, Niger is the highest with a rate of seven children per woman.

Initially, it may seem like having fewer children resulting in a smaller population may be a good thing as it will force humanity to reign in its overgrazing and return to a more stable carrying capacity, but unfortunately, fewer children will also result in fewer labor inputs which will cripple the ability to maintain the scientific edge we have used to inflate our carrying capacity since the 16th century. Technologically inventing our way out of the entropic abyss relies on a growing and healthy economy with lots of inputs which requires a growing population. This lack of having children is mostly due to urbanization and the increased cost of living arising from a lack of frontiers. Why do people migrate? For opportunity. When people migrated into the western frontier they made more humans. When people migrate into cities they have fewer children. Not all opportunities are equal. Over the course of the next 20 years, the mass retirement of humanity's largest working population will rust the machinery of globalism and slow the growth we have all become so accustomed to, and so reliant on. We are already feeling the pains of contraction as more and more humans migrate into cities instead of frontiers as they have done since the dawn of man. This system -the one around you- all of the institutions, and all the mechanisms whereby the modern world operates are all reliant on growth.

Some argue we can adapt and formulate a "post-growth" system, but listen Peter Zeihan (or whatever your name is), history is very clear that there is no such thing as a *good* post-growth system just as there is no such thing as a *good* famine. Scarcity is scary. Post-growth systems are deeply conservative and allow no room for a growing middle class because that would require... growth. All civilizations' twilight years are examples of the post-growth system manifesting, but we do not call these post-growth systems...we call them *Dark Ages*.

The Roman empire made its wealth expanding into frontiers. But lacking combustion engines, computers, AGM-114 Hellfire missiles, and super heavy lift launch vehicles the empire eventually ran out of new frontiers, or at least frontiers that were profitable, see hadrian's wall. When it could no longer expand, towards the end, Roman coins were continuously minted as taxation only met 80% of the imperial budget. This shortfall was met by putting more money into circulation. Since the Roman currency was backed by gold and silver, which have limited quantities, the Roman government couldn't do like the Federal Reserve and just print as much money as they wanted, so to squeeze as many coins out of the metals they had they debased the coins by reducing the valuable metal content. Nero reduced the gold content in Roman coins by nearly 5% and silver by 11%. Commodus, Septimius Severus, and Caracalla all debased Roman coins even more causing more inflation. Gradually, silver coins went from pure 100% silver to 50% silver, and then on down until they reached an all-time low of just 2% silver content. Such blatant manipulation of currency did not go unnoticed by the population who retaliated by paying their taxes using the newer shit coins and keeping the older more valuable ones stored under their mattress.

As the quantity of the Roman empire increased so too did the quantity of resources it had to address problems. But when the quantity of the Roman empire stalled, the quantity of problems demanding resources did not. Additionally, as the quantity of Roman wealth increased the quality of Roman leaders decreased, which caused the quality of the Roman empire to fall, which caused an even greater quantitative increase in the number of resource-requiring problems, which caused the baseless Roman leaders to debase their currency to increase its quantity, causing its quality to fall. This meant the standards were reduced and the ability to produce counterfeit coins was increased, which meant the quantity of counterfeit coins increased to the point that tax collectors were overwhelmed by the flood of fake coins; so the silver currency collapsed, and the Roman government began to collect its tax, not in roman coins, but in *goods themselves*. With the collapse of the Roman monetary system and the return to... well, basic bartering, regional trade collapsed. For the rich and powerful wealth increasingly became measured not in coinage but in the amount of *that which produces goods*: productive people on productive land. Peasants. The more productive land and people one had the more goods one had the more warriors one could hire with said goods to take more productive land from one's neighbors. Feudalism. And so, for about 300 years this anarchic game of Agario was played in the corpse of Rome as petty lords fought to expand their throne.

If we're being honest though, Rome never fell; it simply became a church. The Papacy and its puppet kings rode into the dark ages on the back of peasants, descending into ignorance and strife as cities crumbled and aqueducts fell into ruin as the remaining engineers and stonemasons built forts, not roads, to protect the parasitic nobility from the masses of people they fed on. Slowly the knowledge of how to build anything but war machines was lost, slowly the light of civilization dwindled, health and medicine disappeared and disease ran rampant; all while the Vatican sat on endless archives of Greco-Roman knowledge. Thousands of years of human creativity and innovation, literally buried underground: locked away and dismissed as heretical, purposefully covered in darkness for ages. For they knew that unearthing the light would challenge the darkness over which they ruled. The flash of doubt caused by the spark of an educated genius might have revealed their true forms, showing to the masses that their power was nothing but an illusion, an imposing shadow extinguished by the smallest flame. Despite this, the light was recovered. Slowly. It took centuries of blood and warfare before the ember of antiquity was unearthed ushering in the Renaissance and the Age of Enlightenment.

All of this is to say we need a new frontier. There are post-growth systems. They aren't good. They are brutal, dark, and ugly. Post-growth systems are autocratic in nature, where a minority of elites fearfully control information flows, brutally hoarding what resources are left, greedily guarding the grain silos. We are headed there now. Developing the Moon will avoid this fate. From dust to dust we live, but from lunar crust we ascend. It is not about conquest. It's about carrying capacity. And it's not too late.

Claw out of the entropic abyss.

REFERENCES

ACS Energy Lett. 2020, 5, 11, 3544–3547

ACS Energy Lett. 2023, 8, 2, 1042–1049

Algeo, Thomas J. (5 February 2012). "The P–T Extinction was a Slow Death". *Astrobiology Magazine.*

Allanore, A., Yin, L., & Sadoway, D. R. (2013). "A new anode material for oxygen evolution in molten oxide electrolysis." *Nature,* 497(7449), 353-356.

Beauchamp, Benoit; Grasby, Stephen E. (September 2012). "Permian lysocline shoaling and ocean acidification along NW Pangea led to carbonate eradication and chert expansion". *Palaeogeography, Palaeoclimatology, Palaeoecology.* 350–352: 73–90.

Bennett, E. A.; Crevecoeur, I.; Viola, B.; et al. (2019). "Morphology of the Denisovan phalanx closer to modern humans than to Neanderthals." *Science Advances.* 5 (9)

Bennett, Jay (12 November 2020). "How SpaceX became NASA's go-to ride into orbit". *National Geographic.*

Benton, M.J. (2005). "When Life Nearly Died: The greatest mass extinction of all time." *London: Thames & Hudson*

Berger, Eric (5 March 2020). "Inside Elon Musk's plan to build one Starship a week and settle Mars". *Ars Technica.*

Berger, Eric (2021). "Liftoff." William Morrow and Company. ISBN 978-0-06-297997-1.

Boston Metal. (2023). *Boston Metal.* https://www.bostonmetal.com/

Carson, M. A., Rouen, M. N., Lutz, C. C., & McBarron, J. W. II. (2007). "Biomedical Results of Apollo – Section VI – Chapter 6 – Extravehicular Mobility Unit." *Lyndon B. Johnson Space Center.*

Chandrayaan-3 LIBS confirms the presence of Sulphur (S) on the lunar surface through unambiguous in-situ measurements (August, 2023) www.isro.gov.in

Chapman, P. K. (2017, February). "Losing the geomagnetic shield: A critical issue for space settlement." *NSS Space Settlement Journal.*

Civilini, F., Weber, R., & Husker, A. (2023). Thermal Moonquake Characterization and Cataloging Using Frequency-Based Algorithms and Stochastic Gradient Descent. *Journal of Geophysical Research: Planets,* 128(9)

Cochelin, Anne-Sophie B.; Mysak, Lawrence A.; Wang, Zhaomin (December 2006), "Simulation of long-term future climate changes with the green McGill paleoclimate model: the next glacial inception", *Climatic Change,* 79 (3–4): 381

Dediu, D.; Levinson, S. C. (2018). "Neanderthal language revisited: not only us." *Current Opinion in Behavioral Sciences.* 21: 49–55.

Delfini, Massimo; Kustatscher, Evelyn; Lavezzi, Fabrizio; Bernardi, Massimo (July 2021). "The End-Permian Mass Extinction: Nature's Revolution". *Nature through Time.* Springer Textbooks in Earth Sciences, Geography and Environment. Springer Cham. pp. 253–267.

Dietzler, D. (2016, September). "Making it on the moon: Bootstrapping lunar industry." *NSS Space Settlement Journal.*

Dixit, M. (2023, January). "How space radiation threatens lunar exploration." *Supercluster.*

Dolgova, O.; Lao, O. (2018). "Evolutionary and medical consequences of archaic introgression into modern human genomes". *Genes.* 9 (7): 358

Douglas, M. M.; Douglas, J. M. (2016). "Exploring Human Biology in the Laboratory." Morton Publishing Company. p. 324

Dreibus, G.; Spettel, B.; Wänke, H. Lithium and Halogens in Lunar Samples. Philos. Trans. R. Soc. Lond. A 1977, 285 (1327), 49–54, DOI: 10.1098/rsta.1977.0042

Ellis, J.; Schramm, D. N. (1993). "Could a nearby supernova explosion have caused a mass extinction?". *Proceedings of the National Academy of Sciences of the United States of America.* 92 (1): Pages 235–8

Erwin, D.H. (1990). "The End-Permian Mass Extinction". *Annual Review of Ecology, Evolution, and Systematics.* 21: 69–91.

FAA General Aviation and Air Taxi Activity (and Avionics) Surveys. 2001-2010, 2012-2016,

Farnsworth, A., Lo, Y.T.E., Valdes, P.J., & others. (2023). Climate extremes likely to drive land mammal extinction during next supercontinent assembly. *Nat. Geosci.*, 16, 901–908.

Feng, J., Siegler, M. A., Su, Y., Ding, C., & Giannakis, I. (2023). Layered Structures in the Upper Several Hundred Meters of the Moon Along the Chang'E-4 Rover's First 1,000-m Traverse. *Journal of Geophysical Research: Planets*, 128(8)

Fernández-Armesto, Felipe (2010). "Columbus on Himself." *Hackett Publishing.* p. 270.

Fields, Brian D. (February 2004), "Live radioisotopes as signatures of nearby supernovae", *New Astronomy Reviews,* 48 (1–4): 119–23

Firestone, R. B. (July 2014). "Observation of 23 Supernovae That Exploded <300 pc from Earth during the past 300 kyr". *The Astrophysical Journal.* 789 (1): Page 11.

Geology of the Moon. (2023, September 23). *In Wikipedia.*
https://en.wikipedia.org/wiki/Geology_of_the_Moon

Gläser P., Scholten F., De Rosa D., Marco Figuera R., Oberst J., Mazarico E, Neumann G.A., Robinson M.S. (2014) "Illumination conditions at the lunar south pole using high resolution Digital Terrain Models from LOLA" *Icarus, Volume 243, 2014, Pages 78-90.*

Globus, A. (2012, November). "Paths to space settlement." *NSS Space Settlement Journal.*

Globus, A., & Marotta, T. (2018, September). "How small of a free-space settlement can people be happy living in?" *NSS Space Settlement Journal.*

"Great Dying" lasted 200,000 years". *National Geographic.* 23 (November 2011)

Hans Wedepohl, K. The Composition of the Continental Crust. Geochim. Cosmochim. Acta 1995, 59 (7), 1217– 1232, DOI: 10.1016/0016-7037(95)00038-2

Haskin, L.; Warren, P. Lunar Chemistry. In Lunar Sourcebook, A User's Guide to the Moon; Heiken, G. H., Vaniman, D. T., French, B. M., Eds.; Cambridge University Press, 1991; pp 357– 474.

Hayne, P. O., Aharonson, O., & Schörghofer, N. (2021). "Micro cold traps on the Moon." *Nature Astronomy,* 5(2), 169-175.

Haynes, W. M.; Lide, D. R.; Bruno, T. J. CRC Handbook of Chemistry and Physics; CRC Press, 2016.

Haberl, Helmut; et al. (July 2007), "Quantifying and mapping the human appropriation of net primary production in earth's terrestrial ecosystems", *Proceedings of the National Academy of Sciences of the United States of America,* 104 (31): 12942–47

Heiken, G. H., Vaniman, D. T., French, B. M., Eds. Lunar Sourcebook, A User's Guide to the Moon; Cambridge University Press, 1991.

Heilig, B. (2023). "Lunar modulations." *Nature Physics,* 19(4), 467-468.

Honniball, C. I., Lucey, P. G., Li, S., Shenoy, S., Orlando, T. M., Hibbitts, C. A., Hurley, D. M., & Farrell, W. M. (2021). "Molecular water detected on the sunlit Moon by SOFIA." *Nature Astronomy,* 5(2), 121-127.

Hu, J., Li, S., Liu, H., & Hu, D. (2023). "Reliability and lifetime estimation of bioregenerative life support system based on 370-day closed human experiment of lunar palace 1 and Monte Carlo simulation." *Acta Astronautica,* 202, 609-616. https://doi.org/10.1016/j.actaastro.2022.11.021

ILSA listens to the movements around the landing site (August, 2023) www.isro.gov.in

Industrial Lithium-Ion Battery Emergency Response Guide. (2022, November 11). [PDF file].

"Introducing Megapack: Utility-Scale Energy Storage". (July, 2019) *www.tesla.com*

Jacobs, Zenobia; Li, Bo; Shunkov, Michael V.; Kozlikin, Maxim B.; et al. (January 2019). "Timing of archaic hominin occupation of Denisova Cave in southern Siberia". Nature. 565 (7741): 594–599

Jin, Y. G.; Wang, Y.; Wang, W.; Shang, Q. H.; Cao, C. Q.; Erwin, D. H. (21 July 2000). "Pattern of marine mass extinction near the Permian–Triassic boundary in south China". *Science.* 289 (5478): 432–436.

Jouault, Corentin; Nel, André; Perrichot, Vincent; Legendre, Frédéric; Condamine, Fabien L. (6 December 2011). "Multiple drivers and lineage-specific insect extinctions during the Permo-Triassic". *Nature Communications.* 13 (1)

Jurikova, Hana; Gutjahr, Marcus; Wallmann, Klaus; Flögel, Sascha; Liebetrau, Volker; Posenato, Renato; et al. (November 2020). "Permian–Triassic mass extinction pulses driven by major marine carbon cycle perturbations". *Nature Geoscience.* 13 (11): 745–750.

Krause, J.; Fu, Q.; Good, J. M.; Viola, B.; et al. (2010). "The complete mitochondrial DNA genome of an unknown hominin from southern Siberia". *Nature.* 464 (7290): 894–897.

Labandeira, Conrad (1 January 2005), "The fossil record of insect extinction: New approaches and future directions", *American Entomologist,* 51: 14–29

Lambert, F. (2019, July 29). "Tesla launches its Megapack, a new massive 3 MWh energy storage product." *Electrek.*

Lang, Kenneth (2011). "The Cambridge Guide to the Solar System" (2 ed.). *New York: Cambridge University Press.* p. 199.

Launius, R. D. (2003). "Public Opinion Polls and Perceptions of US Human Spaceflight." *Space Policy,* Volume 19, Issue 3, Pages 163-175

Lavery, Brian (2013). "The Conquest of the Ocean." *New York, NY: DK Publishing.* p. 70.

Lee-Jones, Sarah (September 22, 2021). "New Tesla Megafactory Breaks Ground in Lathrop, California". *Tesla North.*

Li, S., Poppe, A.R., Orlando, T.M., & others. (2023). Formation of lunar surface water associated with high-energy electrons in Earth's magnetotail. *Nat Astron.*

Lunar resources. (2023, October 1). In Wikipedia. https://en.wikipedia.org/wiki/Lunar_resources

Lunar Resources. (2023). *Lunar Resources.* https://www.lunarresources.space/

Lunine, Jonathan I. (2009), "Titan as an analog of Earth's past and future", *European Physical Journal Web of Conferences*, 1: 267–74

Maclean, Frances (January 2008). "The Lost Fort of Columbus". *Smithsonian Magazine.*

"Making Humans a Multiplanetary Species" (September 2016), *SpaceX*. [PDF file].

Marotta, T. (2016, December). "A survey of space settlement designs." *NSS Space Settlement Journal.*

Marshall, Charles R. (5 January 2023). "Forty years later: The status of the "Big Five" mass extinctions". *Cambridge Prisms: Extinction.* 1: 1–13

McDowell, Jonathan (18 May 2022). "Starlink Launch Statistics". *Planet4589.*

Morison, Samuel Eliot (1991) [1942]. "Admiral of the Ocean Sea: A Life of Christopher Columbus." *Boston: Little, Brown and Company.*

M. Smith et al., "The Artemis Program: An Overview of NASA's Activities to Return Humans to the Moon," 2020 *IEEE Aerospace Conference*, Big Sky, MT, USA, 2020, pp. 1-10

Multi-mission radioisotope thermoelectric generator. (2023, September 27). In Wikipedia. https://en.wikipedia.org/wiki/Multi-mission_radioisotope_therm oelectric_generator

NASA Global Climate Change, (June 2021) "Carbon Dioxide Concentration | NASA Global Climate Change", *Climate Change: Vital Signs of the Planet*

NASA Johnson Space Center Office of Communications. (April, 2023). "NASA Successfully Extracts Oxygen from Lunar Soil Simulant." www.NASA.gov

Neron de Surgy, O.; Laskar, J. (February 1997), "On the long term evolution of the spin of the Earth", *Astronomy and Astrophysics,* 318: 975–89

Nishiaki, Yoshihiro; Jöris, Olaf (November 27, 2019). "Learning Among Neanderthals and Palaeolithic Modern Humans: Archaeological Evidence." *Springer Nature.* p. 19.

Novacek, M. J.; Cleland, E. E. (May 2001), "The current biodiversity extinction event: scenarios for mitigation and recovery", *Proceedings of the National Academy of Sciences of the United States of America*, 98 (10): 5466–70

Nuclear power in space. (2023, May 16). In Wikipedia. https://en.wikipedia.org/wiki/Nuclear_power_in_space

Odenwald, S., & Geyer, A. (2011, July). "Radiation Math: Mathematical problems featuring radiation effects applications." *NASA.*

O'Malley-James, J. T.; Greaves, J. S.; Raven, J. A.; Cockell, C. S. (2013), "Swansong Biospheres: Refuges for life and novel microbial biospheres on terrestrial planets near the end of their habitable lifetimes", *International Journal of Astrobiology*, 12 (2): 99–112

Rask, J., Vercoutere, W., Navarro, B. J., & Krause, A. (2017). "Space Faring: The Radiation Challenge: Radiation Educator Guide." *NASA.*

Reich, D. (2018). "Who We Are and How We Got Here." *Oxford University Press.* p. 53

Ruzic, N. P. (1965). "The case for going to the moon." *Putnam.*

Sadoway, D., Ignatiev, A., Curreri, P., & Carol, E. (2019, July). "Regolith extraction through molten regolith electrolysis." *Lunar ISRU 2019-Developing a New Space Economy Through Lunar Resources and Their Utilization*, 2152, 5012.

Sahney, S.; Benton, M.J. (2008). "Recovery from the most profound mass extinction of all time". *Proceedings of the Royal Society.* B. 275 (1636): 759–765

Sawyer, S.; Renaud, G.; Viola, B.; Hublin, J.-J.; et al. (2015). "Nuclear and mitochondrial DNA sequences from two Denisovan individuals". *Proceedings of the National Academy of Sciences.* 112 (51): 15696–700.

Schröder, K.-P.; Connon Smith, Robert (2008), "Distant future of the Sun and Earth revisited", *Monthly Notices of the Royal Astronomical Society*, 386 (1): 155–63

Slon, V.; Viola, B.; Renaud, G.; Gansauge, M.-T.; et al. (2017). "A fourth Denisovan individual." *Science Advances.* 3 (7)

"SOFIA Science Center". (2023) Universities Space Research Association.

Sowers, G. (February, 2020). Thermal Mining of Ices on Cold Solar System Bodies: NIAC Phase I Final Report. Colorado School of Mines.

Soilleux, R. J. (2017, July). "Towards closed environmental control and life support for space habitats Part I: A basic system." *NSS Space Settlement Journal.*

Soilleux, R. J. (2017, July). Towards closed environmental control and life support for space habitats Part II: Reduced risk and increased efficiency with biological systems. *NSS Space Settlement Journal.*

Soilleux, R. J. (2018, January). "Environmental control and life support (ECLSS) for large orbital habitats: Ventilation for heat and water transport and management." *NSS Space Settlement Journal.*

SpaceX Starship. (2023, October 28). In Wikipedia. https://en.wikipedia.org/wiki/SpaceX_Starship

SpaceX. (2023, October 27). In Wikipedia.
  https://en.wikipedia.org/wiki/SpaceX

Specht, Joshua; Stockland, Etienne (2017). "The Columbian Exchange."
  CRC Press. p. 23

St. Fleur, Nicholas (16 February 2017). "After Earth's worst mass
  extinction, life rebounded rapidly, fossils suggest". *The New York
  Times.*

Stanley, Steven M. (18 October 2016). "Estimates of the magnitudes of
  major marine mass extinctions in earth history". *Proceedings of
  the National Academy of Sciences of the United States of
  America.* 113 (42)

Statista Market Insights, www.statista.com

Steigerwald, B., Weibezahn, J., Slowik, M., von Hirschhausen, C. (2023).
  "Uncertainties in estimating production costs of Future Nuclear
  Technologies: A model-based analysis of small modular
  reactors." *Energy*, 281, Pages 128–204.

Stratospheric Observatory for Infrared Astronomy. (2023, October 4). In
  Wikipedia.
  https://en.wikipedia.org/wiki/Stratospheric_Observatory_for_Infr
  ared_Astronomy

Strickland, J., & others. (2019). "National Space Societies Roadmap to
  Space Settlement" (3rd ed.). *Ad Astra Magazine.*

Tammann, G. A.; et al. (June 1994), "The Galactic supernova rate", *The
  Astrophysical Journal Supplement Series*, 92 (2): 487–93

Thomas, K. S., & McMann, H. J. (2011). "US spacesuits." *Springer
  Science & Business Media.* P.120

U.S. Bureau of Labor Statistics. (2023). U.S. Bureau of Labor Statistics.
  https://www.bls.gov/

U.S. Department of Agriculture, Agricultural Research Service. (2023).
  FoodData Central. https://fdc.nal.usda.gov

U.S. Energy Information Administration. (2023). "How much electricity
  does an American home use?"
  https://www.eia.gov/tools/faqs/faq.php?id=97&t=3

U.S. Office of Nuclear Energy. (2019, July 9; updated 2023, June).
  "TRISO particles: The most robust nuclear fuel on Earth."
  https://www.energy.gov/ne/articles/triso-particles-most-robust-nu
  clear-fuel-earth

Waksman, J. (2020, March). "Project Pele overview: Mobile nuclear
  power for future DoD needs." https://gain.inl.gov

Wilford, John Noble (27 August 1985). "Columbus's Lost Town: New
  Evidence Found". *The New York Times.*

Zhang, S., Wimmer-Schweingruber, R. F., Yu, J., & others. (2020). "First measurements of the radiation dose on the lunar surface." *Science Advances,* 6(39).

I invite you to validate any other claims, data, or information contained within this work via the use of state-of-the-art search engine analysis.

ABOUT THE AUTHOR

Ian Long is the creator of *Anthrofuturism* and hopes this work speaks for itself.